GREAT DISCOVERIES IN SCIENCE

Cell Theory:
The Structure and Function of Cells

Carol Hand

Cavendish Square

Published in 2019 by Cavendish Square Publishing, LLC
243 5th Avenue, Suite 136, New York, NY 10016

Copyright © 2019 by Cavendish Square Publishing, LLC

First Edition

No part of this publication may be reproduced, stored in a retrieval system, or transmitted in any form or by any means—electronic, mechanical, photocopying, recording, or otherwise—without the prior permission of the copyright owner. Request for permission should be addressed to Permissions, Cavendish Square Publishing, 243 5th Avenue, Suite 136, New York, NY 10016. Tel (877) 980-4450; fax (877) 980-4454.

Website: cavendishsq.com

This publication represents the opinions and views of the author based on his or her personal experience, knowledge, and research. The information in this book serves as a general guide only. The author and publisher have used their best efforts in preparing this book and disclaim liability rising directly or indirectly from the use and application of this book.

All websites were available and accurate when this book was sent to press.

Library of Congress Cataloging-in-Publication Data

Names: Hand, Carol, 1945- author.
Title: Cell theory : the structure and function of cells / Carol Hand.
Description: New York : Cavendish Square, 2019. | Series: Great discoveries in science | Audience: Grade 9-12. | Includes bibliographical references and index.
Identifiers: LCCN 2018013786 (print) | LCCN 2018017206 (ebook) | ISBN 9781502643704 (ebook) | ISBN 9781502643803 (library bound) | ISBN 9781502643926 (pbk.)
Subjects: LCSH: Cells—Juvenile literature.
Classification: LCC QH582.5 (ebook) | LCC QH582.5 .H36 2019 (print) | DDC 571.6—dc23
LC record available at https://lccn.loc.gov/2018013786

Editorial Director: David McNamara
Editor: Jodyanne Benson
Copy Editor: Michele Suchomel-Casey
Associate Art Director: Alan Sliwinski
Designer: Christina Shults
Production Coordinator: Karol Szymczuk
Photo Research: J8 Media

The photographs in this book are used by permission and through the courtesy of: Cover Lonely/Shutterstock.com; p. 4 Captivelight/Shutterstck.com; p. 10 De Agostini/Biblioteca Ambrosiana/De Agostini Picture Library/Getty Images; p. 16 Science History Images/Alamy Stock Photo; pp. 19, 55 Universal Images Group/Getty Images; pp. 23, 35, 58 Science & Society Picture Library/Getty Images; p. 24 Biophoto Associates/Science Source/Getty Images; p. 27 Yale Joel/The LIFE Picture Collection/Getty Images; p. 28 Andrew Brookes/Cultura/Getty Images; p. 30 Maurice Savage/Alamy Stock Photo; p. 39 Ondrej Korinek/Wikimedia Commons/File:Pomnik J. E. Purkyne (Nove Mesto)Karlovo nam.jpg/CC BY-SA 3.0; p. 42 Ducu59us/Shutterstock.com; p. 45 NYPL/Science Source/Getty Images; p. 52 Science History Images/Alamy Stock Photo; p. 63 Time Life Picture Collection/Getty Images; p. 64 NLM/Science Source/Getty Images; p. 68 Hulton Archive/Getty Images; p. 72 Steve Gschmeissner/Science Photo Library/Brand X Pctures/Getty Images; p. 76 RasmussenImages/Alamy Stock Photo; p. 79 Sebastian Kaulitzki/Shutterstock.com; p. 82 BSIP/UIG/Getty Images; p. 85 Alila Medical Media/Shutterstock.com; p. 92 Sakurra/Shutterstock.com; p. 94 Anyaivanova/Shutterstock.com; p. 97 Kateryna Kon/Shutterstock.com; p. 104 Chris Henderson/Corbis Documentary/Getty Images; p. 109 Molekuul/Shutterstock.com.

Printed in the United States of America

Contents

Introduction 5

Chapter 1: The Problem of Cells 11

Chapter 2: The Science of Cell Theory 31

Chapter 3: The Major Players in Cell Theory 53

Chapter 4: The Discovery of Cell Theory 73

Chapter 5: The Influence of Cell Theory Today 95

Chronology 110

Glossary 114

Further Information 118

Bibliography 120

Index 124

About the Author 128

Before microscopes, biologists could only study the whole body of animals and did not know they were composed of cells.

CELL THEORY

Introduction

In the 1600s, scientists had no way to answer questions about the origins of life. Where did living things come from? How did they reproduce? What were they made of? Did they consist of separate pieces? If so, what did those pieces look like? In those days, there was no way to look closely at living organisms. Microscopes were unknown, so scientists had no idea that microorganisms even existed. They were limited to observing whole plants and animals with the naked eye.

So, the earliest biologists studied whole organisms. Their observations led to development of classification systems. In the fourth century BCE, the Greek philosopher Aristotle divided the living world into plants and animals. He further divided animals into three groups based on movement: walking (land), flying (air), and swimming (water). Aristotle's classification system was based more on behavior than on structure and function, and some animals fit into more than one category. His system lasted until the 1700s.

It was replaced by a system developed by the Swedish botanist Carolus Linnaeus, also known as Carl von Linné. Linnaeus separated life into plant and animal kingdoms and further separated groups of plants or animals into genera (the plural form of "genus") and species, based on their structure. He introduced the system we still use today, known as binomial nomenclature, which identifies each type of organism by a two-part name consisting of its genus and species. Thus, all humans are *Homo sapiens*, and all wolves are *Canis lupus*.

SPECULATIONS on LIFE, PRE-CELL

Despite advances in classification of visible plants and animals, scientists had no way of obtaining evidence about the microscopic details of life. They made up for this by speculating. They concocted explanations for organism structure, function, and reproduction based on prevailing hypotheses about the nature of life. Today, these explanations seem bizarre, but they lasted for hundreds, even thousands, of years.

One of the most widespread concepts was the idea of spontaneous generation, also called abiogenesis, which states that life appears spontaneously from nonliving matter. Abiogenesis was accepted as far back as the fourth century BCE, when Aristotle listed it as one of four methods of reproduction. (The other three were sexual reproduction with copulation, sexual reproduction without copulation, and asexual reproduction.)

A recipe for creating life in the 1600s called for a person to place sweaty underwear and wheat husks in an open bucket and let it sit for several weeks. It was said that during this time, sweat from the underwear would seep into the wheat husks and change them into mice. According to the theory of spontaneous generation, snakes could spontaneously form from horse hairs left in water, maggots formed from rotting meat, and mushrooms sprouted from dead trees.

To describe how the human body functioned, early Greek scientists proposed that four humors, or vital fluids, existed in the bloodstream. They associated the humors with health and disease, particularly nutrition, growth, and metabolism. Each humor was said to serve one of the four elements: blood was associated with air, phlegm with water, yellow bile with fire, and black bile with earth. A proper balance of the four humors was considered necessary to maintain health.

Another popular concept was vitalism, which divided matter into two classes: organic and inorganic. Organic (living) substances were said to contain a "vital force" that inorganic substances lacked. This concept was first proposed during the fifteenth and sixteenth centuries, to counter the view of the French philosopher René Descartes. Descartes had proposed a mechanistic theory to explain life. He declared that animals, including humans, were "automata," mechanical devices that differed from nonliving devices only because they were more complex.

MICROSCOPES and MICROTOMES

The major limitation to studying life up until the 1600s was that scientists were unable to view it up close. This problem was solved with the invention of the microscope. The earliest microscopes appeared in the mid-1600s. They used glass lenses to magnify small objects, and although they were relatively crude and had low resolutions, these instruments opened up a whole new world. From blood flow through a fish tail, to empty chambers in cork cells, to tiny swimming "animalcules" in pond water, seventeenth-century scientists experimenting with the first microscopes discovered objects, organisms, and structures they had previously never imagined, much less seen.

A related invention was the microtome, which sliced animal tissues into very thin slices. Light could be passed through these slices so they could be viewed under a microscope. Before the late 1700s, when the first microtomes were invented, scientists had sliced tissues with razor blades. The resulting slices were imprecise and uneven. The combination of microscope and microtome made it possible to observe the tiny details of organisms.

With these new inventions, scientists observed cells in both plants and animals, and they eventually came to realize that cells were the building blocks of all living organisms. Still, many early users of the microscope tried to fit their observations to currently held ideas, such as spontaneous generation. It took more than two hundred years after the introduction of the microscope before cell theory replaced spontaneous generation as the explanation for the origin and reproduction of living things.

As scientific knowledge slowly built up, leading to a veritable explosion of observations and experimentation, the process of scientific experimentation also evolved. From ancient Greeks, such as Plato, who believed all knowledge was gained through pure reasoning without measurement, the modern concepts of observation, measurement, peer review of results, and controlled experimentation gradually developed. This was true in the biological sciences as well as fields such as physics and astronomy.

Eventually, all serious biologists accepted the tenets of cell theory. Over the past two centuries, this understanding of cell structure and function has expanded into the burgeoning field of cell and molecular biology, which is now a cornerstone of today's biological, genetic, medical, and agricultural sciences. In the next chapter, we will explore the questions that led scientists to discover cells and the mystery that surrounds the nature of cells.

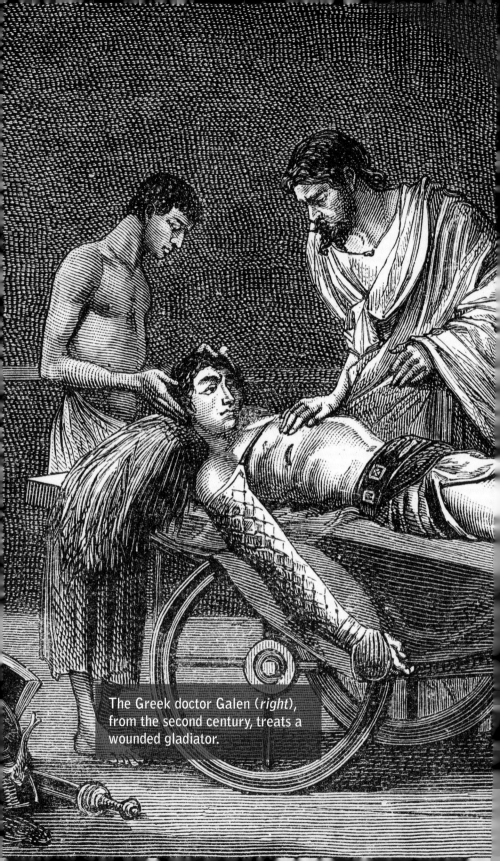
The Greek doctor Galen (*right*), from the second century, treats a wounded gladiator.

CHAPTER 1

The Problem of Cells

From the earliest days of science, educated people wondered about the structure and function of life. But they did not wonder about cells because they had no way of knowing cells existed. Instead, they wondered what was inside the human body that determined health and disease. They wondered how living things differed from nonliving things. They wondered how life regenerated and reproduced.

Scientists in different centuries developed explanations for these problems based on what was known at the time. Because not much was known, those early explanations seem very strange. But some of them became established wisdom and lasted for hundreds, even thousands, of years.

The PROBLEM of HEALTH and DISEASE

The ancient Greeks were extremely interested in human health and disease. Three ancient Greeks who contributed greatly to the early science of medicine were Hippocrates,

Plato, and Aristotle, all of whom lived in the fifth and fourth centuries BCE.

These three men believed health was determined by the four humors, or body fluids, which affected not only physical, but psychological health. They considered blood in animals and sap in plants to be responsible for life. When other body fluids, such as phlegm, bile, and feces, increased in amount, the body's equilibrium was disturbed and disease resulted. For example, they explained the disease epilepsy as phlegm blocking the airways, causing the body to convulse to free itself. They thought health required keeping the four humors at equilibrium within the body.

The Greek doctrine of the four humors described blood, the sanguine humor, as hot and wet. It represented health, vitality, and growth. Its home was the blood vessels, and it carried the "vital force" and "innate heat," which powered cellular metabolism. The phlegmatic humor, considered cold and wet, consisted of all the body's clear fluids—phlegm, mucus, saliva, plasma, lymph, and others. These fluids were said to cool, nourish, lubricate, and protect the organism and to flush out impurities. Yellow bile, the choleric humor, was considered hot and dry. Produced by the liver, its effects were to digest, consume, metabolize, and transform. It was associated with the digestive tract. Finally, black bile, the melancholic humor, was cold and dry, and it had an astringent, precipitating, solidifying effect on metabolism. It held materials in the digestive system long enough to process them properly; in the bloodstream, it allowed blood to clot.

The concept of the four humors may predate even ancient Greece, but it was central to the teachings of the early doctors Hippocrates (for whom the Hippocratic oath is named) and Galen. It remained important in European medicine well into the 1800s.

The PROBLEM of LIFE vs. NONLIFE

Another early biological belief concerned the vital force. This idea existed in ancient Greece, but it became prevalent beginning in the 1600s. The belief in vital force, which attempted to explain the problem of life versus nonlife, was called the theory of vitalism. It arose in opposition to philosopher René Descartes's description of living things as "automata," distinguishable from nonlife only by their complexity. Vitalism assumed that living material contained a vital force not present in nonliving material.

Early descriptions of vitalism simply assumed the existence of a vital force, without attempting to describe it. Later scientists tried to define the concept scientifically. For example, they tested a substance's response to heat. Organic (living) material changed form when heated and could not be returned to its original form. Inorganic (nonliving) material could be melted but could be recovered when the heat source was removed. The assumption was that, once the vital force of a substance was lost, it could not be regained. Vitalists thought this distinction constituted evidence that a vital force existed in living things.

French physiologist Xavier Bichat, a proponent of vitalism, identified twenty-one different tissues in living organisms and explained their behaviors based on these tissues. He assigned living tissues "vital properties" such as sensibility and contractility. He assumed that any substance or object without these properties was not alive.

However, in 1828, the chemist Friedrich Wöhler synthesized urea, an organic compound found in urine, from two inorganic compounds. In the 1800s, this was equivalent to creating life from nonlife. Although it was not Wöhler's intention, his experiment became one piece of evidence against vitalism. Many similar experiments led to the decline of vitalism as a biological concept.

The RISE and FALL of SPONTANEOUS GENERATION

Just as past scientists were not sure how to define or explain life, they were also not sure how life came into being or how it reproduced. But people as far back as ancient Greece accepted the concept of spontaneous generation. This hypothesis stated that the vital force in organic, or living, matter was able to create new living organisms from dead, or inanimate, objects. As evidence of spontaneous generation, they offered examples such as mice "created" from wheat husks mixed with dirty underwear, maggots from rotting meat left in the open, or fish from the mud of a once-dry lake bed. People were particularly open to the idea of spontaneous generation because it fit with their religious views. The vital force that resulted in spontaneous generation, they said, proved God's presence in the world.

Although this concept may seem strange in the modern world, it took nearly two hundred years to finally disprove it. The process of disproving it was doubly important because it marked the beginning of the modern age of experimental science. A series of classic experiments, some of the first to use controls, were done over parts of three centuries.

Redi's and Needham's Experiments

Italian scientist Francesco Redi, in 1668, made the first attempt to disprove spontaneous generation. In a simple but elegant experiment, he placed fresh meat in each of two jars. He left one jar open and covered the other with cheesecloth. Several days later, the meat in the open jar contained maggots, while the meat in the covered jar did not. (There were maggots on the outside of the cheesecloth, however.) When the cover was removed, flies entered the jar and laid eggs, and maggots later hatched. Redi had demonstrated that maggots came from fly eggs, not from rotting meat. This was a problem for spontaneous generation, but even Redi continued to believe it still occurred in some cases.

Englishman John Needham was not convinced. In 1745, he performed an experiment to challenge Redi's experiment. Because it was known that heat would kill living organisms, Needham reasoned that heating chicken broth would kill any organisms in it. He heated broth in a jar, sealed the jar, and let it cool. After several days, the broth turned cloudy,

Francesco Redi did an early experiment designed to disprove spontaneous generation—the idea that life comes from non-life.

indicating that life was present in the jar. Needham concluded from his experiment that life had been created from nonlife, proving the reality of spontaneous generation.

Spallanzani's Experiment

In 1768, Lazzaro Spallanzani, an Italian abbot and scientist, analyzed the results of the Redi and Needham experiments. He questioned Needham's experimental design. Needham might not have heated the broth sufficiently to kill all organisms, or, alternatively, microbes could have entered the flask after the broth was boiled and before the flask was sealed. Spallanzani set up another experiment, placing broth in two bottles and boiling both. He sealed one bottle and left the other open. The sealed bottle produced no life. The open bottle developed many tiny living things, which Spallanzani observed with a recent invention—the microscope.

Spallanzani, like Redi before him, thought he had disproven spontaneous generation. But the debate continued. Some scientists said Spallanzani had deprived the closed bottle of air, which was necessary for the vital force to work.

While scientists argued and experimented, perhaps the person who benefited most from Spallanzani's work was a French chef named Nicolas Appert. The French emperor Napoleon was having trouble feeding his troops. He sponsored a contest to find a better way to preserve food. Appert entered the contest and spent about fifteen years experimenting with food preservation. He copied Spallanzani's techniques, placing food in containers and then boiling and corking it. He began selling the canned food. In 1810, he was awarded the French government's prize, and he is now considered the founder of the canning industry.

Pasteur and the Decline of Spontaneous Generation

Finally, the French scientist Louis Pasteur laid to rest the concept of spontaneous generation. In 1859, he entered a contest sponsored by the French Academy of Sciences, which called for an experiment to prove or disprove spontaneous generation.

Pasteur built on Needham's and Spallanzani's experiments. First, he boiled meat broth in a long-necked flask. Then, he heated the neck of the flask and drew it out into a long S-shape. He reasoned that air (with its dust and organisms) would enter the neck, but gravity would trap dust and organisms in the lowest part of the S-bend, and they would not enter the broth. His reasoning was correct. Broth in the S-shaped flask remained free of bacterial growth, but when the tube was broken and open to the air, bacterial growth occurred rapidly. Pasteur had refuted spontaneous generation and shown that microorganisms live even in air.

In some ways, Pasteur (and earlier scientists such as Redi) were lucky. Some bacteria produce endospores, which are not killed by boiling; therefore, their experiments could have shown growth, as Needham's did. In 1877, John Tyndall published his method for fractional sterilization, which showed the existence of these heat-resistant endospores.

Louis Pasteur used this S-bend flask, which prevents contamination, in his studies to disprove spontaneous generation.

The SCIENTIFIC METHOD GROWS UP

The experiments by Redi, Needham, Spallanzani, and Pasteur (and others in between) illustrate the maturation of the scientific method. The scientists carefully designed experiments to test a hypothesis: spontaneous generation does not occur. They set up conditions to eliminate other possibilities. Boiling the broth (they assumed) eliminated the presence of life. They used controls. Redi's open flask was the natural or "controlled" condition. His closed (experimental) flask showed the result if outside life (such as that of flies) was eliminated. When Pasteur bent the neck of his flask, he prevented airborne organisms from entering the broth (the experimental condition). By later breaking the neck, he allowed air in, setting up the control condition.

All of these scientists wrote about their work. They described their thought processes, experiments, and results. Although communication was much slower in those days, they read each other's work. They practiced peer review. That is, they analyzed others' experiments, tested them to verify the results, and performed their own—hopefully better—experiments, helping to accumulate new scientific knowledge.

Pasteur's experiment—the final in this series—essentially killed the concept of spontaneous generation. Life did not come from nonlife. So where did it come from? The developing technology of the microscope, occurring alongside the experiments in spontaneous generation, would be the key to solving that mystery. It would show the internal structure of organisms, including the presence of cells.

MICROSCOPES: BIOLOGICAL GAME-CHANGERS

A lens is a curved piece of glass, thicker in the middle and thinner around the edges. When ground and polished, it magnifies whatever is seen through it. In the first century, Romans investigated the use of lenses to magnify objects. Centuries later, magnifying glasses were used to view very small insects, such as fleas. These early "flea glasses" had approximately eight- to ten-times magnification.

Around 1590, Dutch spectacle makers Zacharias Jansen and his father, Hans, discovered that by separating two or more lenses within a tube, they could magnify viewed objects. The distance between the lenses determined the magnification. One of their instruments consisted of three sliding tubes. When fully extended to 18 inches (45.7 centimeters), its magnification was nine times; when fully closed, it was three times. Although the lenses were blurry, the Jansens had made the first true microscope. Today, a typical microscope is called a compound microscope because it contains at least two lenses. The second lens further magnifies an image already magnified by the first. It may also be called a light microscope because specimens are viewed through visible light.

But magnification is useless unless there is good resolution—that is, unless the resulting image is clear, not blurry. This requires a balance between focal length (distance between the lens and the point at which the light focuses) and lens diameter. Two scientists in the late 1600s handled this problem in different ways. Antonie van

The Microscopes of Hooke and Leeuwenhoek

Antonie van Leeuwenhoek (1632–1723), a Dutch textile merchant, used tiny glass pearls (spheres of glass) as magnifiers to examine the thread density and quality of cloth. In the 1670s, he began to make these tiny single-lens magnifiers to study the natural world. He made more than five hundred lenses during his lifetime—some as tiny as 1 millimeter (0.04 inch) across. The smaller the lens, the more it could magnify. His lenses magnified objects as much as two hundred to three hundred times.

Leeuwenhoek's minute microscopes were held up to the eye and were difficult to use. He never divulged how he made them. It is likely that he heated the center of a glass rod and pulled the ends outward, forming a long, thin thread. When the end of the thread broke, he placed one end back in the flame until it formed a tiny ball, which he polished to form the final lens. He also never explained how he lit objects for viewing.

Englishman Robert Hooke (1635–1703) made major improvements to the compound microscope. Hooke added a stage for viewing the specimen, an illuminator that shined light through it, and controls to move the lens and focus the specimen. Hooke's microscope could only magnify objects

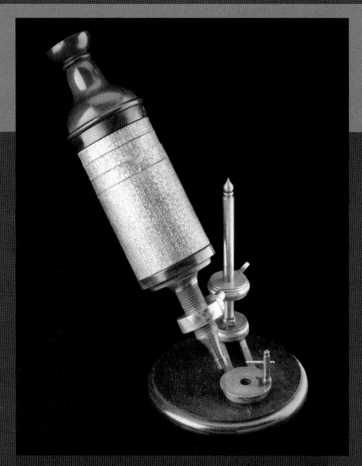

In the 1660s, English scientist Robert Hooke invented one of the first compound microscopes, shown here.

thirty to fifty times. The images were distorted, and the edges were blurry. Even so, Hooke used this improved microscope to view cork slices, leading him to coin the term "cell." He was also the first to use the configuration of three lenses still used in today's microscopes.

Leeuwenhoek made extremely small single lenses from glass balls. His lenses attained very high magnifications, but they were difficult to use.

Robert Hooke used the multiple-lens approach of the Jansens, making a compound microscope. He invented new ways to control the microscope's height and angle. He also varied the light source and direction. Although his

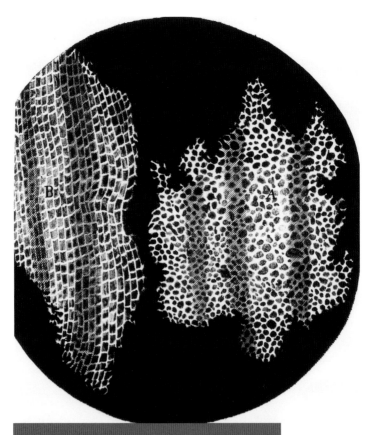

When Robert Hooke viewed thinly sliced cork under his microscope, he saw cells for the first time.

highest magnifications were lower than Leeuwenhoek's (a maximum of fifty times compared to three hundred times), his improvements made the microscope larger and easier to use. Hooke's basic design—much improved by new technology—is still used today.

WHAT HOOKE SAW

In 1665, Robert Hooke published his book *Micrographia*, showcasing his own drawings of observations made with his microscope. Many drawings are close-ups of insects, including a whole flea, insect wings, and insect eyes. But he is most remembered for his close-up of cork cells, which he prepared by slicing a piece of cork very thinly. The cork showed a huge number of empty spaces (1,259,712,000 per cubic inch, by Hooke's calculation), which were pores surrounded by cell walls, left after the cork tree had died.

Hooke did not stop with cork cells. He looked at cells in the plants from his garden and nearby fields—carrots, fennel, elder, burdock, ferns, and Queen Anne's lace, among others. He saw that living plant cells were connected to each other and filled with fluid. But because his microscope was relatively crude, and he was unfamiliar with preparation of specimens, he was unable to learn much more. Still, Robert Hooke gets credit for discovering and naming cells. He also made lasting improvements to the compound microscope, and his book *Micrographia* became the first scientific best seller.

WHAT LEEUWENHOEK SAW

Antonie van Leeuwenhoek was much more interested in what microscopes revealed about life than in the mechanics of the microscope itself. He was familiar with Robert Hooke's *Micrographia*, and he began his work by repeating some of Hooke's observations. For example, his early work includes drawings of bee stings, a fungus, and a human louse.

Then he branched out, making his own discoveries, including the 1674 discovery of single-celled organisms. He called these organisms "animalcules," and we call them protists. Leeuwenhoek was sending letters containing drawings and descriptions of his discoveries to the Royal Society in London, but no one believed his microscopic organisms really existed—until Robert Hooke himself verified them in 1677.

Also, in 1674, Leeuwenhoek viewed and drew red blood cells. Dutchman Jan Swammerdam had first discovered red blood cells in 1668, but Leeuwenhoek described them more clearly and was the first to accurately determine their size. In 1677, he discovered spermatozoa, and later he concluded that they fertilized eggs.

One of Leeuwenhoek's most astounding discoveries occurred in 1676, when he saw bacteria in a drop of water. He estimated it would take ten thousand of these minute organisms to make up the volume of a grain of sand. They were visible only at the limits of magnification of his microscope, and no one else had microscopes so precise. Thus, it was a full century before his observations of bacteria were verified.

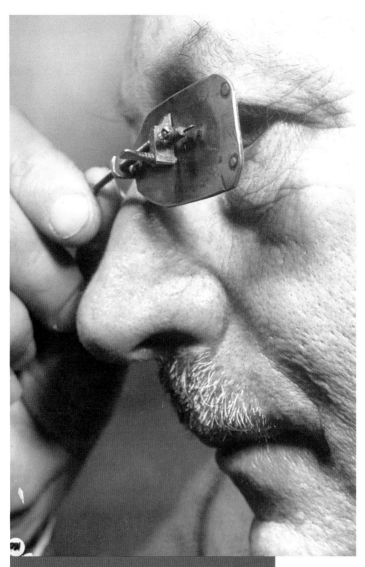

Antonie van Leeuwenhoek's tiny microscopes had to be held up close to the viewer's eye, as shown here.

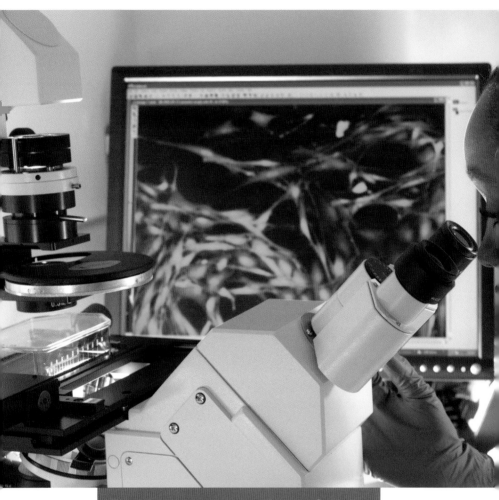

The microscope is essential to the field of biology. This researcher uses an inverted microscope to view stem cells.

MICROSCOPES and CELL THEORY

Although the microscope became one of the keys to the development of modern biology, objects seen under the microscope were still open to interpretation. For example, many early biologists used their microscopic observations to bolster belief in spontaneous generation. They argued that, to obtain tiny organisms, or "animalcules," for observation, they merely had to place a handful of hay in water and let it sit for a few days. The water would soon be filled with "spontaneously generated" organisms.

But although Robert Hooke, Antonie van Leeuwenhoek, and their contemporaries did not always understand what they saw and did not develop cell theory, they did lay the groundwork. They pioneered the use of the microscope to view living organisms and their components. They discovered the hidden world of microorganisms and described many of them. They were the first to see and describe cells in what we now know are multicellular organisms. As technology was improving during this century, it gave scientists better tools for observation—tools that extended their own senses. Microscopes, in particular, opened up a magical, previously hidden world, which would ultimately lead to cell theory and an understanding of how cells and organisms functioned.

Barthélemy Dumortier, a Belgian botanist who contributed to the development of cell theory, was also a prominent politician.

CHAPTER 2
The Science of Cell Theory

Dr. Jaime Tanner, a biology professor at Marlboro College, in Vermont, says a scientific theory is similar to a basket in which scientists keep facts and observations. As they continue to add more facts to the basket, its shape may change. "For example," she says, "we have ample evidence of traits in populations becoming more or less common over time (evolution), so evolution is a fact but the overarching theories about evolution, the way that we think all of the facts go together, might change as new observations of evolution are made."

CELL THEORY as SCIENTIFIC THEORY

A scientific theory is not an opinion, a guess, or a hunch. It is an explanation, a framework that explains all evidence. This evidence includes the observations and experimental results—that is, the facts pertaining to a specific scientific subject. Because a theory must explain *all* known evidence, it must be tweaked to fit any new evidence that is uncovered. The evidence, or the facts, do not change; the theory probably will (at least slightly), as new evidence accumulates. With more evidence, the theory becomes more precise, more robust (stronger), and less likely to change.

Scientific theories describe people's changing understanding of the world. They highlight what we know at a given time about how natural laws work, and they change as scientists discover and understand more. In 1600, scientists did not know that cells existed. In the early 1800s, they were not sure what cells were or what they did. But throughout that century, scientists studied and described cells and observed how they worked. Now, cells are understood to be the smallest, most basic, units of life. This understanding, summarized in cell theory, is essential to our growing understanding of how life works and to all advances in biology and medicine.

The first statements of cell theory, published in 1838 and 1839, were the following:

- All living things are made of cells.

- The cell is the basic unit of life.

- All cells come from preexisting cells.

These statements, plus several others added since, have become a cornerstone of biology. All biologists now accept these statements as a fundamental part of our understanding of life. But they were not always obvious. Even after these concepts were first stated, scientists argued about their truth and what they actually meant. Before these statements could be accepted as scientific theory, what facts or evidence were required?

Scientists first had to understand what a cell was. They had to recognize that cells comprised all living organisms (or, in the case of single-celled organisms, that the cell was the organism). They had to recognize that, while cells varied in size, appearance, and function, all cells contained similar internal structures. They had to recognize that every cell contained all the materials necessary to maintain and reproduce itself and that when cells reproduced, they created more cells just like themselves. In addition, they had to understand, at least partially, how cells did these things. This was a tall order, and it did not happen overnight. It required the work of many biologists, chemists, and other scientists researching and sharing evidence over decades.

PREPPING CELLS for VIEWING

By the late 1600s, biologists were describing the vast and previously unseen living world revealed by the microscope. Instead of guessing what went on inside organisms, they were now viewing it. Instead of arguing about the likelihood of spontaneous generation, they were beginning to figure out how organisms really do reproduce. They were beginning to replace guesswork with detailed observations and controlled experiments.

The late 1600s in cell biology are best known for the contributions of Robert Hooke and Antonie van Leeuwenhoek. Around the same time, Hooke's colleague Nehemiah Grew and Italian physiologist Marcello Malpighi also studied plant cells in detail. Both described the cells' internal structures. Grew hypothesized (incorrectly) that the spaces in plant cells might be similar to gas bubbles in rising bread. Malpighi also published observations of frog lung capillaries and drawings of chick embryo development. Animal cells were harder to study than plant cells because they lacked the thick cell walls of plants and it was more difficult to prepare their thin tissues for observation. But nearly all scientists during this time felt that both plant and animal tissues were composed of some fundamental or "atomistic" unit, and their observations were directed toward finding this unit.

It wasn't until the 1800s that cell science went into high gear. As microscopes continued to improve, new techniques were needed to prepare specimens for viewing under the microscope. They had to be thin enough for light to shine through them, and they needed enough contrast to make their components easily visible. Robert Hooke had viewed fresh tissues, which he sliced with razor blades. But to see cells and their structures in more detail, scientists developed new techniques for preparing plant and animal tissues for viewing. These are called histotechniques (named for histology, the study of tissues). The major tool for tissue preparation was the microtome, which sliced tissues into very thin, even slices.

When preparing tissues, the tissue is first preserved, or fixed (often with formalin), to prevent decay and change. Then, it is embedded in a small block of paraffin and sliced very thinly with a microtome. The microtome holds the paraffin block in a tiny vise while moving it across

This early microtome uses a flywheel and conveyor belt (*right*) to turn a paraffin block for cutting preserved tissues.

the edge of a stationary, very sharp knife. The slicing produces a ribbon of uniform, barely connected sections of tissue-containing paraffin. These are placed in strips on a glass slide for viewing. In modern microtomes, slices can be as thin as one micron (one micrometer, or one millionth of a meter). Often the sliced tissues are stained with special dyes to make the details of their structures stand out.

DUTROCHET STUDIES CELL FUNCTION

A number of scientists had contrasted the different appearances of plant and animal cells. In 1824, Frenchman Henri Milne-Edwards, for example, described animal cells as a mass of globules of uniform size. But the lack of uniformity in actual cells led people to question his observations, and his hypothesis was short-lived.

Henri Dutrochet, another Frenchman working around the same time, had a much more profound influence on cell biology. Dutrochet, a physiologist, wanted to disprove the concept of vitalism and replace it with anatomical and mechanical explanations for the activities of living tissues. His work ranged widely, concentrating on plant physiology but also branching into animal studies. He emphasized the similarities in function between plant and animal cells. He saw the cell as both a structural and a physiological unit, observing, "It is clear that it constitutes the basic unit of the organized state; indeed, everything is ultimately derived from the cell." This is perhaps the first explicit statement of one basic tenet of cell theory.

Dutrochet concentrated on living processes common to both plants and animals, including respiration. In 1832, he showed that the stomata (tiny pores) in leaves let in air, allowing the air to penetrate into deeper tissues. He observed that only plant cells (and only their green parts) can take in carbon dioxide and use it to transform light into chemical energy—an early description of photosynthesis. He tried to show that excitability and motility used the same mechanisms in plants and animals.

Dutrochet also made an important first step in the study of osmosis and diffusion. He showed that some membranes allow water to pass through but block other substances, and that when concentrations differ on the two sides of the membrane, water passes from lesser to greater concentration of dissolved molecules. He even suggested that cells might arise from preexisting cells, another part of the cell theory.

But many of Dutrochet's statements were apparently based on intuition and thought rather than actual observations. He was working with poor tools, and some of his conclusions about similarities between plants and animals were incorrect. His major contribution was to show that similar physiological processes occur in both plants and animals, and these processes can be explained by chemistry—in other words, there is no need to assume the existence of a vital force.

PURKINJE COMPARES ANIMALS and PLANTS

Czech doctor and physiologist Jan Evangelista Purkinje (or Purkyňe) lived from 1787 to 1869. He studied many cells and tissues of the human body. Several body structures are even named for him, among them the Purkinje cells of the brain's cerebellum. He studied, among other things, blood vessels in the retina, muscle contraction, effects of various drugs on humans, nerve cells in mammals, human skin and sweat glands, and the structures of bone, teeth, cartilage, arteries, and veins. He compared cell structures in plants and animals. He even developed a microscopy course at his university, the University of Breslau, Prussia, teaching students the tools and techniques necessary to study tissues under the microscope.

Several of Purkinje's studies were directly applicable to cell theory. He studied chicken eggs before they hatched and coined the term "protoplasm" (meaning "first form") to describe the gelatinous substance inside the egg. He discovered the egg cell, or ovum, within the egg. He corresponded with German scientist Ernst Karl von Baer, who later discovered ova in mammals. In 1837, Purkinje stated that animals were composed primarily of cells and cell products (but thought fibers might also be a component) and that animals' "basic cellular tissue is … clearly analogous to that of plants." These statements, his observations on egg cells, and his descriptions of "granules" in animal and plant cells and the protoplasm in developing cells were studied by Theodor Schwann in Germany. Schwann was also studying cell structure. He was one of two scientists who later received most of the credit for the original cell theory. (He did not give credit to Purkinje.)

CELLS HAVE ORGANELLES

Scientists quickly noticed that cells were not uniform structures. Their gel-like protoplasm contained darker inclusions, later called organelles, or tiny organs. Observations of the most obvious organelles in these early years laid the groundwork for understanding the way cells work. They led to the conclusion that each organelle was specialized to perform a specific cell function. Showing that cells functioned like tiny self-contained units, maintaining and reproducing themselves, made it much harder to believe in old concepts such as the four humors.

Czech physiologist Jan Purkinje was highly honored for his many biological discoveries, some relating to cell theory.

In 1665, Robert Hooke discovered the structure surrounding the cell, now known as the cell membrane. The next important discovery was the liquid-filled cell vacuole, which is very large and obvious in living plant cells. It was first described by Antonie van Leeuwenhoek in 1676.

The most visible organelle in most cells is the nucleus. Czech botanist Franz Bauer first described the nucleus in 1802. However, Bauer did not give the structure a name. The term "nucleus" was coined by Scottish botanist Robert Brown, who first saw nuclei as opaque brown spots in the epidermis of orchid cells. Brown observed nuclei in hundreds of plants, many of them during a research trip to Australia. He soon recognized that nuclei were present in other cell types, including developing pollen grains. However, he did not realize that animal cells also had nuclei. He described plant cell nuclei in a speech in 1831; the Linnean Society published his paper on the topic in 1833. Because he coined the name and described the organelle in many different plants, Brown is credited as the discoverer of the nucleus, but he gave credit to Bauer.

Discovery and description of new cell organelles continued throughout the 1800s and into the mid-1900s. As each structure was found, scientists worked to determine its function. Two series of discoveries in particular helped set the stage for cell theory by shedding light on two key functions of the cell: cell reproduction and the transformation of energy.

CELLS REPRODUCE by MITOSIS

The understanding of cell reproduction and where cells come from required a series of breakthroughs spanning many years. One of the earliest occurred in 1766, when Abraham Trembley observed bacterial reproduction by the process of binary fission (the division of a single cell into two identical daughter cells).

French biologist François-Vincent Raspail, an expert microscopist, gave excellent descriptions of plant cells that also described their functions—for example, their ability to organize themselves and reproduce. In 1825, Raspail described this process with the phrase "omnis cellula e cellula," or "every cell is derived from a [preexisting] cell." This is one of the three statements in the original cell theory. However, German physician and biologist Rudolf Virchow usually gets credit for this phrase.

In 1832, a Belgian scientist, Barthélemy Dumortier, described what he termed "binary fission" in plants. What Dumortier actually observed was cell division or, more specifically, the formation of a cell plate between new cells at the end of cell division. He reached a conclusion similar to Raspail's conclusion: cells do not come from within old cells, but rather from preexisting cells.

Despite these excellent descriptions of the process, cell reproduction remained controversial. Matthias Schleiden, later credited (along with Theodor Schwann) with the original cell theory, continued to insist that new cells formed

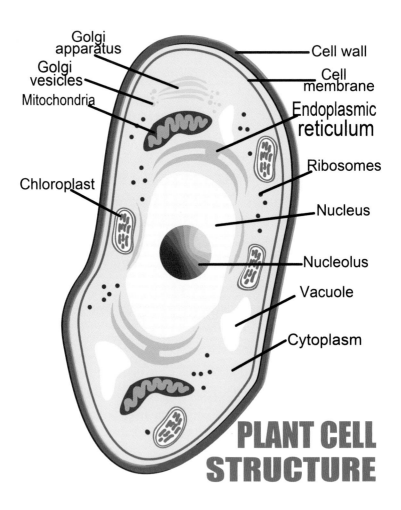

PLANT CELL STRUCTURE

The discovery that the same organelles occurred in all cells helped scientists begin to understand how cells work.

by a process of crystallization that began either inside or outside the cell—in short, a form of spontaneous generation. In 1852, German scientist Robert Remak published his observations on cell division, which disputed Schleiden's explanation. Remak stated that new animal cells reproduced by binary fission (now called mitosis). Here again, the originator of an idea failed to get the credit. Rudolf Virchow widely publicized Remak's observations, and, as with Raspail, he usually receives credit for Remak's observations.

A major piece in the puzzle of cell reproduction was the 1842 discovery of chromosomes by Karl Wilhelm von Naegeli. In Naegeli's essay on pollen formation, he described cell division and stated that the wall that formed down the center of the dividing cell was the result, not the cause, of cell division. He described chromosomes, calling them "transitory cytoblasts," but had no inkling of their function. But Naegeli's studies did convince him that Schleiden's theory of cell reproduction was incorrect.

Walther Flemming, in 1879, introduced the term "mitosis" to describe cell division and described the longitudinal splitting of chromosomes during this process. In 1883, Wilhelm Roux suggested that each chromosome carried a different set of heritable information, which was divided equally when the chromosomes split. This showed how new cells form within organisms. Mitosis is now understood to produce two genetically identical daughter cells and to be the process for growth and repair in all body cells.

SEX CELLS REPRODUCE by MEIOSIS

It was clear from the behavior of chromosomes during cell division that they were essential to the transfer of genetic information from cell to cell. But how did they get from organism to organism, or generation to generation? That is, how did they affect heredity? To understand this, it was logical to look at gametes, or sex cells (ova, or eggs, in females and spermatozoa, or sperm, in males)—the only obvious cellular link between generations. A major question was the location of genetic information. Eggs and sperm were very different in size, but their nuclei were approximately equal in size, so nuclei seemed a good starting place. Chromosomes were the most obvious constituent of the nuclei, and scientists were intrigued that the number of chromosomes remained constant—among cells within an organism, among individuals within a species, and among generations.

Scientists began to wonder how organisms maintained their chromosome number. In other words, how did organisms maintain the same number of chromosomes in their body cells from one generation to the next? This puzzle was solved by a series of discoveries over about seven decades. These discoveries led to an understanding of meiosis, the process of cell division that occurs during the formation of gametes. Body cells have two copies of every chromosome; this is known as the diploid, or 2n, number. Meiosis, or reduction division, results in the formation of gametes having only one set of chromosomes. This is known as the haploid, or n, number. When an egg and a sperm meet and share their DNA, the original diploid chromosome number is restored.

Theodor Schwann, shown here, was one of two scientists credited with the first statement of cell theory.

Giving Credit Where It's Due—or Not

Plagiarism—the act of publishing another person's words, ideas, or data as one's own, without giving credit—is a serious breach of ethics. Anyone who commits plagiarism is discredited.

Plagiarism is usually prevented because scientific papers go through peer review before being accepted for publication. Also, scientists today have easy access to scientific information, through journals and the internet. They work together and constantly share information.

But in the 1800s, two scientists deeply involved in developing cell theory blatantly took credit for others' work. Theodor Schwann discussed his work with his colleague, Matthias Schleiden. When Schwann published his version of cell theory in 1839, he included Schleiden's ideas without giving him credit.

Schwann also studied the work of Jan Purkinje, who stated that plant and animal cells appeared very similar. Although this idea was central to Schwann's contribution to cell theory, he also failed to credit Purkinje.

Rudolph Virchow, who is credited with the third tenet of cell theory, also plagiarized ideas. Although Virchow still receives credit for the statement that cells come from preexisting cells, François-Vincent Raspail had made this statement several years earlier. Virchow also published Robert Remak's ideas on mitosis. But Remak fought back.

Remak had worked in Virchow's laboratory and had published his observations three years earlier. He wrote Virchow, pointing out the similarity of Virchow's ideas to his. Virchow dismissed Remak's objections, but he felt obligated to defend himself. In an 1858 book, he wrote that, because his earlier piece was only an editorial, there was no need to credit Remak. Virchow suffered no consequences for his plagiarism.

Rudolf Albert von Kölliker, in 1840, is credited with first realizing that eggs and sperm are both types of cells. In 1856, Nathanael Pringsheim observed a sperm cell penetrate an egg cell, confirming the process of fertilization. German biologist Oscar Hertwig, in 1876, first observed meiosis in sea urchin cells. In 1883, Belgian zoologist Edouard van Beneden described how chromosomes move during meiosis, this time in *Ascaris* (roundworm) egg cells. Two other biologists, the German August Weismann in 1890 and the American Thomas Hunt Morgan in 1911, explained the importance of meiosis in reproduction and inheritance.

Austrian monk Gregor Mendel, in 1865, had done a detailed study of heredity in the garden pea, but his results were largely ignored. In 1902, two biologists—American Walter Sutton and German Theodor Boveri—independently recognized that the movement of Mendel's "factors" during gamete production in peas was exactly the same as the movement of chromosomes during meiosis. This led to the recognition that genes are located on chromosomes. Sutton and Boveri thus share credit for the development of the chromosome theory of heredity.

CELLS TRANSFORM ENERGY

Another major problem in early cell biology was determining how cells transform energy. Unraveling the answer began with the discovery of two organelles: mitochondria and plastids.

The mitochondrion is now known as the cell's "powerhouse." Mitochondria convert the chemical energy contained in food (that is, in carbohydrates, fats, and proteins) into the chemical adenosine triphosphate (ATP), which can be rapidly broken down by cells during cellular respiration. ATP breakdown provides cells with energy for all life functions; thus, mitochondria are essential in both plant and animal cells.

Like the explanation for cell reproduction, it took many years and many scientists to discover mitochondria and figure out how they transform energy. The first clue was uncovered in 1857, when Rudolf Albert von Kölliker and other scientists began noticing "granules" in muscle and other cells. But their microscopes lacked the resolution to see details of the structures. In 1886, German cytologist Richard Altmann used a dye technique to view them more closely. He called them "bioblasts," and he considered them to be basic units of cellular activity. In 1898, German microbiologist Carl Benda named them mitochondria, from the Greek words for thread (*mitos*) and granule (*chondros*).

Although mitochondria were quickly associated with energy transformations, figuring out how they accomplished this took much longer. Scientists worked through much of the twentieth century to describe their structure and work out the complex series of chemical reactions by which they transform energy. They identified proteins called cytochromes in mitochondria that allowed oxygen to be processed. They isolated ATP and showed its importance in cellular respiration. They showed that oxidation (changes in chemical substances when oxygen is added) occurred within mitochondria. Not until 1978 was the purpose of mitochondria fully established, in part by Nobel Prize winners Peter D. Mitchell and Paul Boyer.

In addition to mitochondria, plant cells also contain chloroplasts, a type of plastid, or pigment-containing cell. Chloroplasts are the sites of photosynthesis, by which plants make their own food. During photosynthesis, chloroplasts in plants take in carbon dioxide and water and transform it (using energy from sunlight) to produce sugars (plant food) and oxygen. Excess sugars are stored in plants by bonding them together to form starch.

Figuring out photosynthesis also involved a long path—more than two hundred years. Two early scientists discovered necessary raw materials for photosynthesis, but neither of them understood their results. In the 1600s, Belgian chemist Jan Baptista van Helmont discovered that plants require water to grow. In the 1700s, Joseph Priestly discovered the presence of oxygen in air, and Dutch scientist Jan Ingenhousz showed that plants produce oxygen in the presence of light. Swiss botanist Jean Senebier, in 1796, showed that plants, in the presence of sunlight, use carbon dioxide and release oxygen.

In the 1860s, Julius Sachs studied the relationship of light and chlorophyll to the production of starch in plants. Another key step involved the discovery of the exact organelle responsible for photosynthesis. Credit for discovering plastids goes jointly to two German biologists: Ernst Haeckel, who coined the term "plastid" in 1866, and Andreas Franz Wilhelm Schimper, who, in 1883, defined the term more precisely and distinguished among various types of plastids. (A plastid is a major double-membrane organelle found in the cells of plants, algae, and some other eukaryotic organisms.) Finally, in the 1930s, Cornelis B. van Niel developed the general equation for photosynthesis.

Over the four hundred years between 1600 and 2000, cell biology gradually evolved. Scientists, through constant observation and testing, discovered the presence of cells in living organisms and their various components and organelles. They gradually teased out both the structures and the chemical reactions involved in cell functions, particularly cell reproduction and energy transformations.

The understanding that cells could create and use organic chemicals, move materials from place to place, and generally maintain themselves eventually discredited old beliefs such as the four humors and the concept of vitalism. The understanding that cells could reproduce themselves through mitosis and transfer information to new generations through meiosis explained organism reproduction, once credited to spontaneous generation. Scientists continued to sharpen their experimental skills, developing testable hypotheses and ultimately sound scientific theories to explain their observations. One of these was cell theory.

When German chemist Friedrich Wöhler, shown here, synthesized the organic compound urea from inorganic components, he helped disprove vitalism.

CHAPTER 3

The Major Players in Cell Theory

Scientific discoveries are never made by a single person. Even if one person has a tremendous breakthrough, that person is building on the work of many who have gone before. Cell theory is no exception. The discovery and improvement of the microscope as well as the thousands of observations and insights of those who used microscopes (plus chemical techniques) to study cells over several centuries all combined to result in the modern-day cell theory.

However, there are always a few whose contributions stand out—either because they were first to state part of the theory, because they did the research that resulted in that part, or sometimes simply because they were given the credit. The modern-day cell theory consists of six tenets, or principles. These tenets were not all developed at once. The first two originated in 1839, the third a few years later, and the last three near the end of the 1800s. It took this long for the combined work of many scientists—the majority of them in Germany—to uncover and agree upon the secrets of cell structure and function.

German scientists Matthias Schleiden and Theodor Schwann are usually given credit for the first two tenets of cell theory. German Rudolph Virchow is credited for the third. These three principles involve the general structure and reproduction of the cell. The last three tenets involve cell physiology—that is, the way cells function to maintain themselves and to reproduce. A number of people receive credit for these. They include the following: Walther Flemming, Wilhelm Roux, Walter Sutton, and Theodor Boveri for tenet four, and Louis Pasteur, Friedrich Wöhler, and Justus von Liebig for tenets five and six. Who were these men? Let's find out.

MATTHIAS SCHLEIDEN

Matthias Schleiden (1804–1881) was born in Hamburg, Germany. He came from a wealthy family and originally went to law school at the University of Heidelberg, in Germany. He practiced law for several years but hated it. After suffering a deep depression, he attempted suicide. In 1831, he gave up the law and enrolled in Jena University, also in Germany, to study botany and medicine, his true interests. After graduation, he remained at the university, becoming a full professor in 1850.

While at Jena University, Schleiden studied plant development. Over his career, he developed improved microscopic techniques and teaching methods, which he introduced in his classrooms and in a botany textbook (*Principles of Scientific Botany*, 1842–1843). His textbook went through several editions, and Schleiden used his new

Matthias Schleiden conducted many years of observations on plant cells to develop his ideas about cell theory.

microscopic techniques to develop a deeper understanding of the plant cell. This marked the beginning of the cell theory and of the science of plant cytology.

Schleiden was the first person to recognize that cells were the fundamental unit of life, at least in plants. (He did not study animals.) This was the vital first component of cell theory. He studied Robert Brown's discovery of the cell nucleus (which he renamed the "cytoblast") and added his own observations. He concluded that every new plant embryo originates from a nucleated cell, and thus the nucleus was essential to cell division. This also became a component of cell theory. However, not all of Schleiden's conclusions were correct. For instance, he thought new cells erupted from the surface of the nucleus like blisters.

Schleiden was not a perfect scientist. For example, he seemed unable to criticize and verify his own hypotheses. His lively imagination and tendency to speculate made other scientists consider him too liberal. But his approach stimulated his students and younger botanists. His classic textbook changed the study of botany from the purely systematic approach used since the days of Carolus Linnaeus (in which botany consisted simply of classifying plants) to a vibrant analysis of plant structure and function. His new approach eventually earned him the title "reformer of scientific botany." But Schleiden was apparently a restless man. After 1850, he left botany and devoted the rest of his life to the study of philosophy and history.

THEODOR SCHWANN

Theodor Schwann (1810–1882) was born in Neuss, Germany. His early education was at a Jesuit college in Cologne, and he remained a devout Roman Catholic his entire life. Beginning at age eighteen, he studied medicine and natural sciences at the University of Bonn, in Germany, under Johannes Müller, the leading physiologist of the 1800s. Schwann became Müller's doctoral student and later his research assistant at the University of Berlin, also in Germany. Müller believed in conducting science through observation, but he still accepted vitalism, the prevailing view at the time. Thus, both Roman Catholicism and vitalism influenced Schwann's early life.

Schwann received his medical degree at age twenty-three and became a physiologist and anatomist. From 1834 to 1838, he worked in Müller's lab, conducting experiments that provided data for Müller's classic textbook, *Elements of Physiology*. In Schwann's experiments, he disproved spontaneous generation, discovered the gastric enzyme pepsin, and identified the role of yeast in converting sugar to alcohol (the process of fermentation). His fermentation data were ridiculed until Louis Pasteur confirmed the process a decade later.

Schwann's most important accomplishment during this time came from his collaboration with his colleague, Matthias Schleiden. In 1838, Schleiden had published his most famous paper, in which he made a partial statement

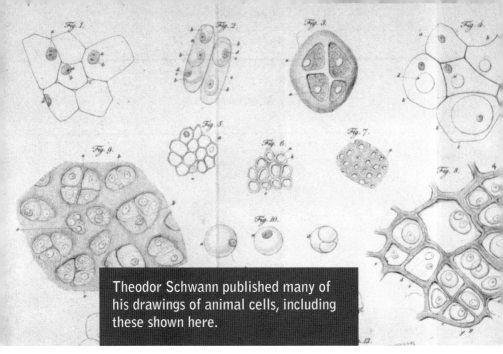

Theodor Schwann published many of his drawings of animal cells, including these shown here.

of the original cell theory. One evening, as the colleagues were socializing, they began discussing their work on cells. When Schwann heard Schleiden's descriptions of plant cell nuclei, he realized that the animal cells he had observed were very similar. Schleiden and Schwann immediately went into Schwann's lab to view his slides and compare the similarities, including the presence of nuclei.

After further study of the structure of peripheral nerve cells, Schwann expanded Schleiden's plant cell theory to include animal cells. He presented his evidence to the Academy in Paris. The following year, in 1839, at the age of twenty-eight, he published his major book, *Microscopical Researches into the Accordance in the Structure and Growth of Animals and Plants*. In this publication, he identified the cell as the basic unit of structure in an organism. He proposed a three-part cell theory, only two parts of which were correct. He also coined the term "metabolic" to describe the chemical processes occurring in living cells and tissues.

Schwann had gone one step further than Schleiden by recognizing that animals, as well as plants, were composed of cells and that all their cells contained a nucleus. But his book failed to acknowledge the contributions of any other scientist, even Schleiden.

Schwann received many honors during his lifetime. In 1845, he received the prestigious Royal Society Copley Medal for his work in cytology. He was appointed professor of anatomy at the University of Liege, Belgium, in 1848, and chair of physiology in 1858. In 1879, he was elected to the Royal Society (London) and to the French Academy of Science. He was an excellent teacher whose students liked him. As Schwann aged, he became more immersed in religious thought. He retired in 1880 and died two years later.

RUDOLF VIRCHOW

Theodor Schwann made one major error in his statement of cell theory. His third tenet, on the notion of how cells form, described a structureless substance that he named "blastema." He concluded that new cells were produced by a type of crystallization or spontaneous generation within this substance. Another German scientist, Rudolf Virchow, received credit for setting this matter straight.

Rudolf Carl Virchow (1821–1902) was born in the town of Schivelbein, in the German kingdom of Prussia (now located in Poland). He received a classical education in high school and eventually became fluent in seven languages in addition to German. In 1839, he received a scholarship to attend the Prussian Military Academy, in preparation for becoming an army physician. However, after receiving his

Schwann and the Cell Theory

> The cause of nutrition and growth resides not in the organism as a whole, but in the separate elementary parts—the cells.
>
> The development of the proposition, that there exists one general principle for the formation of all organic productions, and that this principle is the formation of cells, as well as the conclusions which may be drawn from this proposition, may be described by the term cell-theory.

These quotations from Theodor Schwann's book, *Microscopical Researches into the Accordance in the Structure and Growth of Animals and Plants* (1839), state the essence of his cell theory. But before beginning his discussion of cells, Schwann contrasts two types of thinking: teleological and physical.

Schwann explains teleological thinking as a kind of thinking that presumes a purpose in the creation and function of every organism; that is to say, organisms are designed in a certain way by "a power acting with a definite purpose."

He explains physical thinking, on the other hand, as the view that holds that living (organic) beings are structured in the same way as inorganic matter, based on necessity and the constraints of matter, not with any specific purpose.

Schwann points out that teleological thinking had been discarded in the physical sciences but still held sway in biology. He uses his observations of cells and tissues to argue against the teleological viewpoint.

Today, teleological thinking in biology (or any other science) is dismissed, but in Schwann's day, vitalism was still accepted—for example, by his mentor, Johannes Müller. (Vitalism assumed living material contained a "vital force" that was not present in nonliving material.) Schwann obviously felt a need to prove that his thinking met modern standards.

medical degree, he became a professor in Berlin, where he remained for most of his life.

Virchow's passion was pathology, or the science of diseases, especially the study of diseased tissues in the laboratory. He constantly admonished his students to "think microscopically" and to recognize that diseases originate in specific cells and tissues, not in the whole body. It was this insight that led to his famous statement "omnis cellula e cellula," meaning "all cells come from preexisting cells." Virchow received credit for this phrase, but French biologist François-Vincent Raspail said it first (in 1825). The phrase became the third tenet of cell theory. Virchow's work also made medical diagnosis much more precise and helped launch the field of cellular pathology.

Most of Virchow's ideas about medicine were very modern—for example, he utterly rejected the use of Galen's "humors," an idea that still persisted in medicine. He insisted on evidence-based, well-researched science. His modern beliefs put him at odds with editors of medical journals, who refused to publish some of his papers. Frustrated, he teamed up with another young physician, Benno Reinhardt, and the two founded their own journal. Now named *Virchows Archiv*, it remains a preeminent medical journal.

A highly energetic man, Virchow maintained an active political career in tandem with his pathology research. He was elected to the German Parliament in 1880, representing the German Progress Party, which he helped found. His liberal politics served social justice causes, such as better education and living conditions for the poor. At the same time, he pursued research into the causes of diseases

Rudolf Virchow, shown here, developed part of cell theory and was also an important public health activist.

and developed the field of "social medicine," in which he promoted better health care for the poor. But, despite his modern leanings, he did not get everything right. Sent to investigate a deadly typhus epidemic, Virchow accepted the prevailing medical wisdom and blamed its cause on "miasmas," or bad air caused by rotting matter. He did not recognize that diseases were caused by microorganisms.

WALTHER FLEMMING

The last three tenets of the modern cell theory were based on research into the function of cells—both cellular reproduction and the transformation of energy. In the understanding of cell reproduction, the standouts include at least four people: Walther Flemming, Wilhelm Roux, Theodor Boveri, and Walter Sutton.

Walther Flemming (1843–1905) was born in the town of Sachsenberg, Austria (now located in Germany). He studied medicine at the University of Rostock and worked as a military physician before becoming a professor, first at the

German biologist Walther Flemming pioneered the science of cytogenetics with his description of the process of mitosis.

University of Prague, in the Czech Republic, and later at the University of Kiev, in Ukraine, where he taught anatomy.

While studying cell nuclei in the gills and fins of salamanders, Flemming used aniline dye to make granular materials in the nuclei stand out. Because the granules absorbed the new synthetic red dye, Flemming named the red-stained granules chromatin (Greek for "color"). He discovered that, during mitosis, the chromatin condensed to form thread-like structures. Later, German anatomist Heinrich Waldeyer would name these strands chromosomes. Flemming described the various stages in the process of mitosis, or cell division, including its longitudinal splitting to form two new cells.

Based on his observations, Flemming concluded that each cell nucleus originates from a previously existing cell nucleus, showing how the third tenet of cell theory happened. In 1882, he published his research in a book, *Cell Substance, Nucleus,* and *Cell Division.* His discovery of mitosis is listed as one of history's top one hundred scientific developments and one of the ten most important discoveries in cell biology.

WILHELM ROUX

But, although Flemming showed how chromosomes moved during mitosis, he did not understand their importance in heredity. In 1883, Wilhelm Roux would provide a clue relating to the role of chromosomes in transferring genetic information to a new generation.

Wilhelm Roux (1850–1924) was born in Jena, Germany. Before and after serving in the military during the Franco-Prussian War, he attended the University of Jena. He received his medical degree in 1877. He worked as a professor at the Anatomical Institute in Breslau, becoming an important figure in the field of experimental embryology.

Roux's early research considered the relationship of cell activity and the development of organs in embryos. In 1883, he proposed a new theory regarding the role of chromatin in development. He suggested that chromatin was not uniform but composed of a variety of substances, which resulted in daughter cells having unique characteristics. This was perhaps the first suggestion of how chromosomes were sorted during egg and sperm formation—a cell division process that would come to be known as meiosis.

Based on his experiments with frog embryos, Roux proposed that the information present in daughter cells determined how these cells would develop. When he destroyed one blastomere (cell in a two-cell embryo), the remaining blastomere developed into half an embryo. Roux also suggested that the longitudinal splitting observed by Flemming ensured equal division of the heritable elements in the dividing cells.

Roux continued his embryology research throughout his life as an anatomy professor. He belonged to thirty-seven professional societies and received numerous honors as founder of the first journal of experimental embryology.

THEODOR BOVERI and WALTER SUTTON

Theodor Boveri (1862–1915) was another major player in cell biology. In 1904, he confirmed Flemming's observations and Roux's theory about the longitudinal splitting of chromosomes during division. Like Roux, Boveri was primarily an embryologist, but much of his research revolved around chromosomes and their genetic contribution to the embryo.

Boveri was born in Bamberg, Germany. He was a highly creative and artistic child, and his parents made sure he was well trained in music and art. He continued to draw and paint throughout his scientific career. He considered art essential to the practice of science. He also held his students to his own high work ethic and standards. According to Fritz Baltzer, Boveri's biographer, Boveri was "highly creative and intelligent, yet humble and mature beyond his years." He also made profound contributions to cell theory.

Boveri studied anatomy and botany at the University of Munich, in Germany, receiving a doctorate in 1885. While working at the Zoological Institute in Munich, he read the work of Edouard van Beneden on egg maturation and fertilization in a species of the nematode worm, *Ascaris*. Boveri also began to study *Ascaris*. The fact that it had few chromosomes made it a good research model. Boveri's research repeated and expanded upon that of van Beneden, Flemming, and Roux. He confirmed the role of chromosomes in carrying genetic material and showed that the splitting of chromosomes equally divided the genetic material. He published his results in 1902 and 1903. In his papers, he connected chromosomes to the hereditary "factors" described by Gregor Mendel in his studies on inheritance in the garden pea. Mendel's work, published in 1865, had only recently been rediscovered.

While Boveri studied *Ascaris* in Munich, a young graduate student at Columbia University, in New York City, was coming to the same conclusions, working on grasshopper cells. Walter Sutton (1877–1916), born in Kansas City, had graduated from the University of Kansas and a few years later went to New York City to work with renowned biologist E. B. Wilson in Columbia's Department of Zoology. Sutton made his breakthrough in 1902, when he was only twenty-five. He figured out that chromosomes are the basis of heredity and that chromosome numbers are reduced by half during the formation of sex cells (the process of meiosis). Sutton further realized that the chromosome activity he was observing related directly to

German biologist Theodor Boveri discovered that, during the process of meiosis, the chromosome number is halved.

Mendel's laws of inheritance. In 1902 and 1903, Sutton published papers describing his experiments and drawing conclusions about them.

Boveri and Sutton, working independently, came to exactly the same conclusions regarding chromosomes, their movements during meiosis, and their importance in heredity. Their joint discovery is now known as the Boveri-Sutton chromosome theory. It was an important discovery on its own, but it also became a key piece of cell theory.

Unfortunately, Walter Sutton did not continue his promising work. He also did not live to see the astounding developments in cell theory that would follow. He left research to become a surgeon, serving in France as a physician during World War I. He died at the age of thirty-nine, following an appendectomy.

CHEMICAL COMPOSITION and ENERGY TRANSFORMATIONS

The last two tenets of cell theory get into the details of how cells do what they do. If cells make up living things, reproduce themselves, and get passed from generation to generation, how do they do it? What about their structure and function makes it possible for them to carry out these complex actions? How do they maintain themselves? How do they interact with each other to form an organism? They can do these things because all cells have very similar chemical compositions (tenet five) and because all cells transform energy in the same way—through cellular respiration and (in the case of plants) through photosynthesis (tenet six).

Again, many scientists were responsible for piecing together the information on chemical structure and energy transformations in cells. But a few stand out. One person important in determining the chemistry of cells was Friedrich Wöhler. Others were responsible for identifying the organelles responsible for energy transfer. Rudolf Albert von Kölliker and Carl Benda helped identify the mitochondrion; Julius Sachs, Ernst Haeckel, and Andreas Schimper were instrumental in understanding photosynthesis and identifying plastids.

FRIEDRICH WÖHLER

In the 1800s, scientists began to absorb the fact that cells can maintain themselves without the need to invoke a vital force. Friedrich Wöhler played a key role in proving this fact. Wöhler (1800–1882) was born in Frankfurt, Germany. He received a medical degree from the University of Heidelberg, in Germany. But his first love was chemistry, and he gave up medicine to study with renowned chemist Jons Berzelius in Stockholm, Sweden. The two became close friends. Wöhler made several important achievements in chemistry, including isolation of the element aluminum.

Wöhler's mentor, Berzelius, believed in vitalism. Wöhler decided to test Berzelius's belief that organic compounds could not be made from inorganic reactants. In 1828, when Wöhler heated lead cyanate and ammonia solution, crystals of the organic compound urea were formed, revealing a problem with the idea of vitalism. His success spurred other scientists to try synthesizing organic from inorganic compounds. The theory of vitalism was finally considered dead in 1845, when chemist Adolf Kolbe created the organic compound acetic acid by combining carbon, oxygen, and hydrogen.

Wöhler's position as professor of chemistry at Göttingen University, in Germany, kept him busy for the rest of his life. He taught many students and helped establish one of the first teaching laboratories at the university.

The DISCOVERY of MITOCHONDRIA and PLASTIDS

Because mitochondria were so tiny, it took many years for scientists to realize that they were real organelles and even longer to understand what they did. A number of people participated in the discovery of mitochondria, including Rudolf Albert von Kölliker, Richard Altmann, and Carl Benda. Benda, in 1898, finally named them "mitochondria," or thread granules. It was well into the twentieth century before their role in energy transformations was described.

Kölliker had a role in developing cell theory, but he did not set out to find mitochondria. He was an important physiologist and embryologist, noted for helping confirm the third tenet, which states that cells can come only from preexisting cells. He also supported the study of tissues as groups of individual cells. This helped cement the idea that each cell is an independent, functioning unit.

The understanding of photosynthesis likewise has a long, complicated history. Julius Sachs, in the 1860s, and Cornelis van Niel much later, in the 1930s, made major contributions to understanding the chemical nature of photosynthesis and its relationship to light and chlorophyll. Ernst Haeckel and Andreas Schimper, in the late 1800s, respectively discovered and categorized plastids.

Thus, no one person can be credited with the discovery of the last two tenets of the modern cell theory. By this time, communication among scientists was much faster, laboratories were much larger, and much of scientific discovery involved large groups of scientists working together.

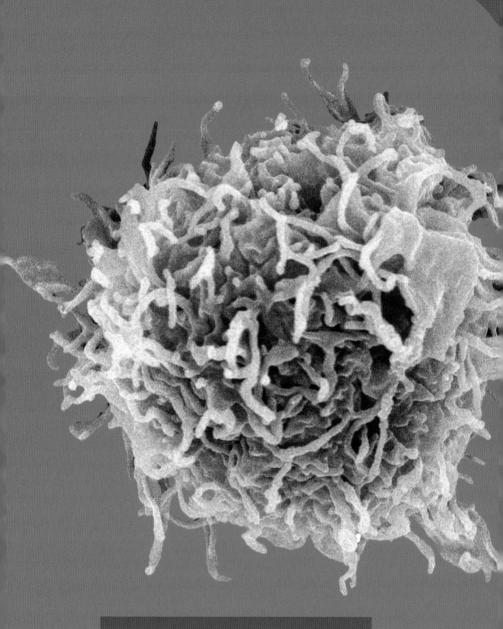

Electron microscopes enable scientists to study cells up close. This white blood cell, a B cell, is involved in immunity.

CHAPTER 4

The Discovery of Cell Theory

For any science to progress, it must have a coherent underlying structure. Otherwise, even if discoveries are made, scientists in that field do not know how to process the information. Dr. Mehdi Tavassoli, of the Scripps Clinic and Research Foundation, points out that this was the situation in the 1600s, when Robert Hooke first observed cork cells. He did not understand what he was seeing, and, according to Tavassoli, Hooke considered the finding "merely a curiosity." It took another two hundred years for scientists to realize that all organisms are made of cells. But when the original cell theory finally emerged, it began what Tavassoli called a "golden age for biology."

The MODERN CELL THEORY

Since the first statements of cell theory in the 1830s and 1840s, cell theory evolved rapidly as techniques and technologies for microscopic viewing improved. In the twenty-first century, it forms a powerful basis underlying the entire field of biology. The modern cell theory varies slightly, depending on who presents it, but it generally includes six tenets:

1. All living organisms are composed of cells.

2. The cell is the basic structural and functional unit of all living things.

3. All cells come from preexisting cells by cell division (mitosis and meiosis).

4. Cells contain heredity information in their DNA; this information is passed to new cells and new generations by cell division.

5. All cells have the same basic chemical composition.

6. All energy flow necessary for life (metabolism and biochemistry) occurs within cells.

These six tenets may seem somewhat overlapping, but they all mean something different. Although it is not usually stated, the last two tenets incorporate the idea that most functions necessary for chemical reactions and energy flow occur within cell organelles. All six tenets have resulted from the accumulated research and insights of many scientists working together and separately on different organisms. Eventually they came to the same inescapable conclusions, which apply to all living organisms. All six were known at least in general by the end of the nineteenth century, but many of the details have been (and continue to be) filled in and bolstered by research in the twentieth and twenty-first centuries.

ALL LIVING THINGS ARE COMPOSED of CELLS

If something is alive, it is composed of cells. That is to say, every living thing—from the tiniest bacterium to the largest blue whale or giant sequoia to every human being on Earth—is composed of one or more cells. A living organism is able to grow, reproduce, respond to stimuli, and adapt to its environment. Because many living organisms are single cells, those single cells must be able to do all these things.

Of course, early scientists did not look at cells from all existing organisms before constructing the cell theory. But they looked at many thousands, and they found that every single type of organism either was a cell or it consisted of cells. They began to understand that their previous assumptions about life—beliefs such as vitalism and the four humors—could not be true. They began to formulate a new framework for understanding life, a framework based on observations and evidence.

Eventually scientists determined that all living things could be classified into one of six kingdoms of life. Two of these kingdoms are types of bacteria—the kingdoms Bacteria (the typical bacteria that always surround us) and Archaea, or ancient bacteria (most of which live in extreme environments, such as hot springs). These organisms are all

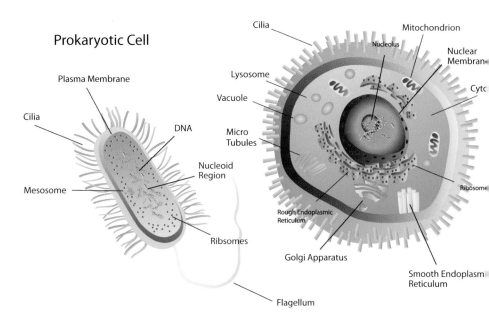

Eukaryotic cells have nuclei and organelles and are much larger than prokaryotic cells such as bacteria.

unicellular, or single-celled. They are known as prokaryotes. "Pro" means "before," and "karyon" means "nut" or "kernel." The nut or kernel refers to the cell nucleus—that is, prokaryotes lack a cell nucleus. Prokaryotic cells are extremely small and simple. Their DNA is not protected by a nucleus; instead, it is free inside the cell. They also lack other complex organelles. They are surrounded by a cell membrane, and most also have a cell wall.

All other living organisms are eukaryotes, having true ("eu") cells. They include animals, plants, fungi, and protists. Their cells are much larger than bacterial cells. They have nuclei, which shelter their DNA, and various other complex organelles, such as mitochondria and chloroplasts, that carry out specific cellular functions. Most of these organelles are membrane-bound—that is, they are composed of or surrounded by a complex membrane with a specific chemical structure. The cells themselves are also membrane-bound. Animals, plants, and fungi are all multicellular, or many-celled. Most protists are unicellular.

A key difference between prokaryotes and eukaryotes is size. Bacterial cells range in size from one to ten micrometers (millionths of a meter); a typical animal or plant cell can range from ten to one hundred micrometers. If a pinhead is half a millimeter in diameter, and a typical *E. coli* bacterium is two micrometers (2 μm) long, about 250 bacteria will fit end to end across the pinhead's diameter. A typical plant (eukaryotic) cell is about fifty micrometers (50 μm) across; only ten of these plant cells will fit across the pinhead.

The CELL IS the BASIC UNIT of ALL LIVING THINGS

It might seem that the first two tenets of cell theory are the same—that is, if all living things are composed of cells, then the cell is the basic unit of all living things. But there is a distinction. The first tenet helps in separating life from nonlife and in classifying types of life. If an organism is a cell, or is composed of cells, it is a living thing, and it can be placed in one of the kingdoms based on its other characteristics.

The second tenet refers primarily to cell structure and function. Cells are the organizing factors in living things. Unicellular organisms are cells. Multicellular organisms are composed of groups of cells. The cells vary in size and shape. They have the same types of organelles and groups of compounds, but these may vary in number and specific type based on their function within the organism. A human body, for example, has nerve cells, muscle cells, and blood cells, among others. These cells are organized into tissues—groups of cells that are similar in structure and perform a similar function. Nerve cells (and a few other types) form the nervous tissue of the brain and spinal cord. Blood consists of several types of blood cells and the liquid plasma.

Groups of different tissues come together to form organs, such as the heart, liver, lungs, or brain. Organs form organ systems—the nervous system, respiratory system, or circulatory system, for example. Finally, all organ systems are connected to form whole multicellular organisms. All of these levels of organization begin with cells, and all

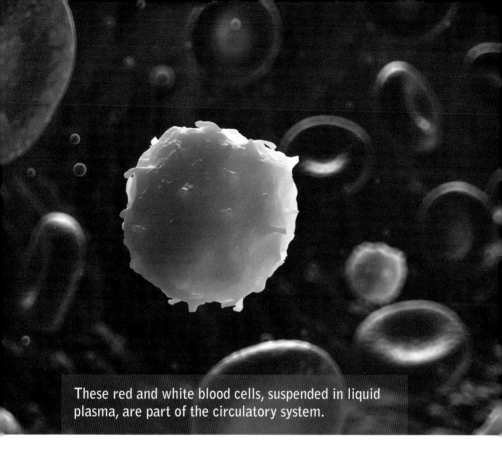

These red and white blood cells, suspended in liquid plasma, are part of the circulatory system.

of the cells work together and communicate to form an interconnected system—the organism.

The functions of living organisms—growth, reproduction, energy transformation, and so on—are all performed within the individual cell. How all the cells work together determines how (and whether) the organism functions. This happens by means of chemical reactions. Many of these chemical reactions occur on the cellular membrane that encloses the cell or on the membranes that enclose or make up other organelles, such as the nucleus and mitochondria. Other reactions occur in the cytoplasm (liquid or gel-like matter that contains the organelles and is enclosed by the cell membrane).

ALL CELLS COME from PREEXISTING CELLS

Acceptance of the third tenet of cell theory overcame the centuries-old problem of spontaneous generation. This tenet states that cells do not come from nothing, from the air, or from rotting meat. They come from cells that already exist. Looked at it in this way, the third tenet says that life is a continuous process. It explains how cells, once formed, continue to divide, forming new generations of cells. It explains all growth in organisms as well as maintenance and repair of cells as they age, wear out, or become injured. For example, as a person sheds skin cells, new skin cells are forming beneath them (by mitosis) to take over. A cut heals as new cells form (by mitosis) to take the place of injured cells.

Formation of new cells involves growth in size and production of new organelles from raw materials taken in as food and present in the cell. But before this can happen, the cell must duplicate its DNA so that a complete copy of the DNA is passed on to every new cell in the next cell generation. DNA is called the genetic material because it is inherited, or passed on to new generations of cells and organisms. DNA is a nucleic acid, a type of organic molecule. (RNA is another example of a nucleic acid.)

Most prokaryotes (bacteria) make new cells, or increase their numbers, by binary fission. In this process, a bacterial cell duplicates its DNA and then divides, producing two identical new cells from the first. Because the DNA of the new (daughter) cells is identical to the parent cell's DNA, the daughter cells are clones of the parent. They are originally smaller than the parent, but they take in nutrients and grow.

When they reach adult size, these new cells can now undergo binary fission themselves. This is asexual reproduction.

In eukaryotic organisms, DNA is located on chromosomes inside the nuclei. Chromosomes are divided into genes. Each gene provides directions for making one specific type of protein in the cell. All the genes together provide directions for making every chemical the cell needs. Together they make up the organism's structure and define its behavior and activities.

A new eukaryotic organism produced by sexual reproduction begins with a fertilized egg, or zygote. A zygote forms when the DNA in an egg combines with the DNA in a sperm. Every new organism formed from a zygote has the same amount of DNA (the same number of chromosomes containing the same number of genes) as its parents had, but the zygote's DNA consists of a mixture of the characteristics of both parents. This makes the new organism unique. It is one reason for the great diversity of life on our planet.

After the zygote is formed, it begins to divide by the process of mitosis. All the many millions of cells in an organism are formed by mitosis. This process was first described by Walther Flemming and others more than a century ago. Scientists have since divided the process into rather arbitrary steps that make it easier to understand. These steps can be observed in preserved cells, such as the onion root tip cells often used in biology classes.

Exactly how many cells do organisms have? That's not easy to figure out. Of course it varies by individual and species, but the calculation has been attempted for humans. The conclusion is that the average human has 37.2 trillion cells. That's a lot of mitosis!

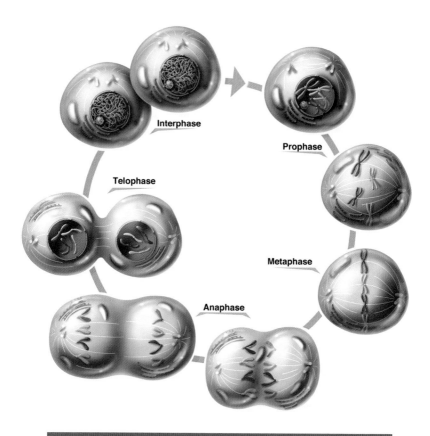

During mitosis, a cell duplicates its chromosomes and then divides, forming two identical daughter cells.

Each mitotic division produces two identical daughter cells. The cells are identical to the parent cell in terms of their DNA content; they have the same number of chromosomes and genes. This means all of the organism's genes are present in every cell. But obviously, all cells in an organism do not look alike. A nerve cell is very different from a skin or liver cell. These differences occur because, shortly after mitosis begins, the developing organism also begins to differentiate, or change its shape and activities. Cells move to new locations within the tiny developing embryo, and depending on where each cell ends up, it is predetermined to become part of the brain, lung, heart, skin, or other organ.

In adult organisms, cells undergo mitosis in place (within their own tissue), and no further differentiation occurs. Old liver cells produce new liver cells, old skin cells produce new skin cells, and so on. This enables the organism to maintain itself, repairing or replacing damaged or diseased cells throughout life.

CELLS CONTAIN HEREDITARY INFORMATION

Mitosis passes DNA on to new generations of cells within an organism. But DNA must also be passed on to new generations—from parent to child—because DNA provides the blueprint for making new organisms. This is the concept encompassed by the fourth tenet of cell theory.

Each species, or type of organism, has its own blueprint, called its genome. For example, the cat genome ensures that cats create more cats, rather than dogs or oak trees, when they reproduce. Slight variations in the DNA (in the genes for eye or fur color, for example) result in differences in the basic pattern. Cats may have blue, green, or yellow eyes; their fur might have a solid, tabby, or Siamese pattern. While a kitten is similar to its parents, it also shows variation because of the mixing of the genes from each parent.

In order to understand the transfer of hereditary information, scientists had to learn how cells could be transferred to the next generation without doubling the chromosome number. This transfer could be explained by the process of meiosis during gamete formation. Information about the movements of chromosomes was revealed by observing egg and sperm cells under light microscopes. Among other things, it showed differences between mitosis and meiosis.

Mitosis involves one DNA duplication followed by a single division into two cells. Meiosis, in contrast, involves one DNA duplication followed by two cell divisions. The first division produces two new cells with their chromosome numbers reduced by half. They are haploid, which refers to having a single set of unpaired chromosomes. The second is like mitosis. It divides each cell again, producing two cells identical to the haploid parent. This produces a total of four cells per parent cell. In males, this results in the development of four sperm. In females, one very large egg cell develops, and the other three are lost. All of the gametes are haploid.

The development of electron microscopes in the 1950s enabled scientists to discover the details of a process during meiosis that makes gametes unique, or unlike their parent cells. Chromosomes in cells are paired. They originally receive one of each chromosome from their father and one from their mother. Before meiosis begins, they duplicate their DNA. Each chromosome now has two segments (called chromatids) connected in the middle. Each chromatid has a complete set of that chromosome's DNA.

Meiosis I

Prophase I

Metaphase I

Anaphase I

Telophase I

Cytokinesis

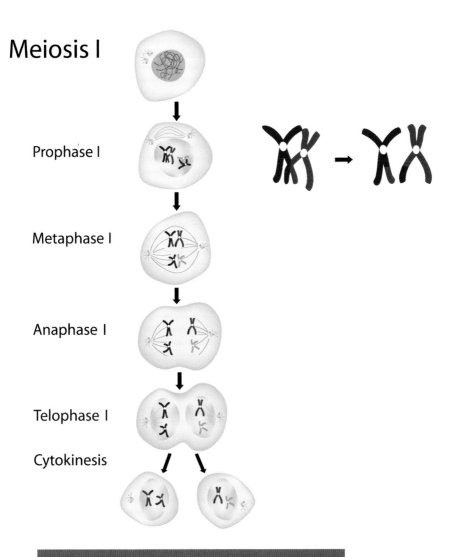

Meiosis I is the first of two divisions in meiosis. When paired chromosomes are aligned (*right*), they sometimes exchange genes.

The Evolution of Microscopes

Microscope technology has made giant strides since the days of Hooke and Leeuwenhoek. Today's standard, the compound light microscope, has an illuminator built into the base. Below the stage is a condenser with lenses that focus light onto the specimen and a diaphragm that controls contrast. Light passes through the specimen and enters one of three or four objective lenses on a rotating structure. Magnifications of objective lenses range from four times to one hundred times. The light then goes through one or two ocular lenses (through which the viewer looks), which magnify ten times to fifteen times. These microscopes have total magnification up to about 1,200 times and resolutions to about 0.25 micrometers. This makes most bacteria and some mitochondria visible.

Some light microscopes are specialized. Phase contrast microscopes enhance contrast in living specimens. Fluorescence microscopes view structures that bind fluorescent dyes. This enables scientists to dye a substance and track it through cells or tissues.

Electron microscopes illuminate specimens using a beam of electrons, rather than visible light. They can achieve resolutions as small as

0.2 nanometers (nm, or billionths of a meter), so specimens are magnified several hundred thousand times or more. The details of objects as small as protein and DNA molecules are visible.

A transmission electron microscope (TEM) passes an electron beam through extremely thin sections of tissues or viruses. An image is produced on a fluorescent screen or photographed and stored on a computer. In a scanning electron microscope (SEM), a beam of electrons travels over a specimen's surface. The electrons produce an image on a monitor, which is photographed or recorded digitally. The SEM has a lower resolution than the TEM but produces three-dimensional images of the specimen's surface.

As they prepare to make their first (reduction) division, the chromosomes line up in pairs down the center of the cell—the mother's and father's contributions side by side. While in this position, the chromatids of adjacent chromosomes sometimes wrap around each other and exchange pieces, a process called recombination. Thus, the chromosomes entering new gametes are not exactly like those in the parent cells. This, combined with the mixing of chromosomes when egg meets sperm, results in offspring with different combinations of characteristics than either parent.

ALL CELLS ARE CHEMICALLY SIMILAR

The fifth tenet of cell theory considers the similarity of chemical composition of cells in all types of organisms. Early scientists recognized the presence of certain gases, such as oxygen and carbon dioxide, in cells and organisms. As the field of chemistry became precise enough to identify compounds, more substances were identified, and it became clear that the same chemical elements and compounds reappeared over and over in the contents of living cells. Elements are the smallest units of any kind of matter. We now know that approximately 95.65 percent of every living organism, by weight, is composed of only four elements: carbon, nitrogen, oxygen, and hydrogen. Adding sulfur and phosphorus brings that value to more than 99 percent.

These six key elements are present in living tissue in approximately the same proportions. They are combined in specific ways to form compounds. Water is a compound composed of two molecules of hydrogen and one molecule of oxygen; hence, its chemical designation is H_2O. Water is vital to living cells; in fact, 99 percent of the molecules in a cell are water molecules, and about 70 percent of a cell's weight is due to water.

There are relatively few types of compounds found in living cells. There are perhaps a thousand different kinds of small organic (carbon-based) molecules. Some of these serve as building blocks to make the much larger organic compounds, such as proteins and nucleic acids, that are vital to cell structure. Others are energy sources used in thousands of metabolic reactions in the cell. Some serve both functions.

These small organic molecules are grouped into four major types: sugars, fatty acids, amino acids, and nucleotides. Sugars are the cell's major energy source. Excess sugars are stored by bonding them together to form huge carbohydrate molecules, such as starch and glycogen. Fatty acids are a building block for lipids, the major energy storage molecule in animals. Lipids are also vital in cell structure and function. The membranes of all cells and the membranes of their organelles are composed of two layers of lipids, with proteins embedded in them. Amino acids are the building blocks of proteins, which are vital as structural components and in nearly all chemical reactions within cells. Many proteins are enzymes, necessary to start or speed up reactions. Finally, nucleotides are the basic structural unit of nucleic acids, including DNA and RNA, which contain and transfer genetic information.

In other words, these four groups of small organic molecules form the building blocks of the four groups of large organic compounds that make up the structure and carry out the functions of cells and, therefore, organisms.

In addition, a cell contains a small percentage of dissolved inorganic molecules, such as gases and minerals. These are used to make compounds or participate in chemical reactions. Waste products from cellular metabolism stay in the blood and tissues for a short time, but in a healthy organism, the cells quickly flush these out of the body.

CELLS CARRY OUT METABOLISM

As a living organism, or part of a living organism, every cell must take in and use energy from the outside environment. The final tenet of cell theory refers to the processes that make this energy flow and utilization possible. All chemical reactions occurring in the cell contribute to the cell's use of energy, or its metabolism.

Green organisms (including plants and algae) can make their own organic compounds for food. They do this by the process of photosynthesis, which makes sugars using carbon dioxide, water, and energy from the sun. Photosynthesis converts light energy into the chemical energy of organic compounds. This process takes place in the cell's chloroplasts, which contain the green pigment chlorophyll.

The electron microscope has given scientists a close-up view of cell organelles. Chloroplasts show an intricate structure consisting of two layers: a relatively smooth and oval outer membrane and an inner membrane (or stoma) containing stacks of oval structures filled with chlorophyll and other light-absorbing pigments. The inner structure also contains proteins, which participate in the many chemical reactions needed to carry out photosynthesis.

Plants store the excess sugars made during photosynthesis as starch. To use the stored energy, they must break down the bonds in the sugar or starch, releasing the energy. They (and all non-photosynthetic organisms) do this through cellular respiration, the process that converts chemical energy into energy for cell activities, including movement, growth, reproduction, and synthesis. Cellular respiration takes place in the cell's mitochondria, the organelles described as the "powerhouse of the cell." The number of mitochondria is proportional to a cell's level of activity. An extremely active heart muscle cell has several thousand mitochondria. A much less active skin cell has fewer.

Like chloroplasts, mitochondria also have a complex structure composed of membranes. The outer membrane is smooth; the inner membrane is highly folded. The folds provide more surface area to hold proteins used in the reactions in the electron transport chain. This series of reactions produces most of the ATP, or energy-containing molecules, during cellular respiration. The inner membrane is also the site of many of the other reactions in the respiration process.

All living organisms—bacteria, fungi, and animals as well as plants and algae—must carry out cellular respiration to obtain energy from organic compounds. But non-photosynthetic organisms, which cannot make their own food by photosynthesis, must eat or otherwise absorb organic compounds in the form of food. The process of cellular respiration is the same in all organisms, regardless of how they originally obtained their food.

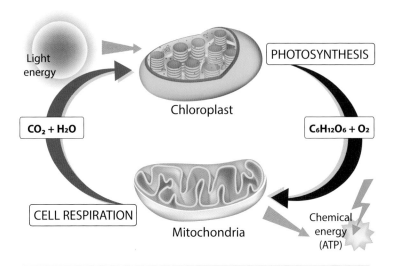

Products of photosynthesis provide reactants for cellular respiration, and vice versa. Photosynthesis stores energy; cellular respiration releases it.

While both photosynthesis and cellular respiration involve several series of complex reactions, the close relationship between them is apparent in their summary equations. The reactants in cellular respiration are the products in photosynthesis and vice versa.

The summary reaction for photosynthesis is:

$$\text{carbon dioxide} + \text{water} \xrightarrow[\text{chlorophyll}]{\text{sunlight}} \text{glucose} + \text{oxygen}$$

The reactants (or raw materials) are carbon dioxide and water. In the presence of sunlight and chlorophyll, these reactants will go through the reactions necessary to produce the products glucose (a plant sugar) and oxygen. Energy is stored in the glucose.

The summary reaction for cellular respiration is:

$$\text{glucose} + \text{oxygen} \longrightarrow \text{carbon dioxide} + \text{water} + \text{ATP}$$

In this equation, glucose is broken down, using oxygen, to release carbon dioxide, water, and ATP energy, which is used for the cell's activities.

In other words, photosynthesis produces the food needed for cell activities, and the products of cellular respiration provide the reactants needed for photosynthesis. Thus, as long as sunlight energy is available, the two processes work together, providing the raw materials to keep both systems going.

These six tenets describing cell theory began with three simple statements first published in 1839. In the nearly two hundred years since then, scientists have continued to refine the theory, answer many of the early questions about cells, and come up with more and more questions. Our understanding of how cells are constructed and how they work continues to grow.

Today's biology students routinely use modern, complex light microscopes to study cells.

CHAPTER 5
The Influence of Cell Theory Today

Acceptance of cell theory led to the development of an entire new scientific discipline: cell and molecular biology. Over the past two centuries, this field has resulted in our understanding of prokaryotic and eukaryotic cells, leading to the modern classification of living kingdoms. It has led to an understanding of the structure and function of the cell membrane and other cell organelles. Combined with advances in chemistry, it led to the identification and understanding of the organic molecules making up living cells, movement of materials across cell membranes, chemical reactions within cells, and the processes of photosynthesis and cellular respiration.

Advances in the late twentieth and early twenty-first centuries have been both theoretical and practical. Basic research is done to expand knowledge, with no specific application in mind—that is, it is theoretical. In almost all cases, basic research later becomes the basis for practical applications, but it begins just because scientists are curious and want to understand something better. Basic research has rapidly increased our understanding of the origin and evolution of cells, how cells work, and how their activities are controlled through DNA. Some significant areas of basic research have been development of the endosymbiotic theory and the genetic code. Applied research results in practical applications of new knowledge. For example, the development of genetic engineering (gene editing and sequencing) techniques is advancing the treatment and cure of many diseases and improving agricultural products. Reproductive technologies, such as artificial insemination, use detailed knowledge of reproductive cells to improve the chances of having a baby. The future promises likely advances in microscopy, live cell imaging, and stem cell research as well as molecular bioengineering (cellular nanotechnology) in areas such as food production and waste disposal.

ENDOSYMBIOTIC THEORY

The endosymbiotic hypothesis—the idea that chloroplasts might have evolved as organelles when photosynthetic bacteria took up residence inside larger cells—was first proposed in 1905. But there was no evidence, and even after a series of experiments in the 1930s, the hypothesis was not taken seriously.

Then, in the 1960s, new microscope techniques enabled scientists to distinguish between prokaryotic and eukaryotic cells. They also discovered that both chloroplasts and mitochondria contain their own DNA. These discoveries led American geneticist Lynn Margulis to revive the endosymbiotic hypothesis. In 1970, she published her theory that mitochondria and chloroplasts had once

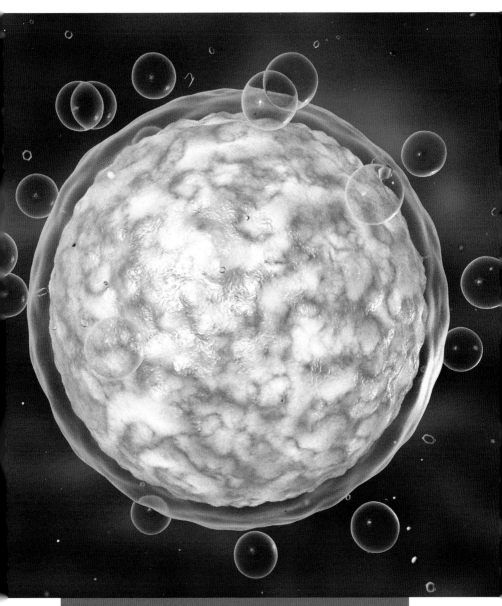

Reproductive technologies involve manipulation of human egg cells, such as this one, to increase couples' chances of having a baby.

The Ethics of Stem Cell Research

Stem cells are rapidly dividing, undifferentiated cells that have the potential to become any type of body cell. All cells in very early embryos are stem cells because they have not yet begun to differentiate.

Adult stem cells are also called somatic stem cells. A few places in the adult human body still produce stem cells but only for the tissue in which they are found. Stem cells in skin, for instance, only produce more skin cells.

Some people consider stem cell research a tremendous boon to humanity, while others consider it immoral. Stem cell research gives scientists tremendous potential to learn about cell growth and development. It also opens the possibility of developing treatments or cures for diseases and injuries as diverse as Alzheimer disease, Huntington disease, multiple sclerosis, diabetes, spinal cord injuries, and some cancers.

But a major objection to stem cell research is the ethical question of destroying human embryos. Stem cells for early research came from the destruction of blastocysts formed from laboratory-fertilized human eggs. Those not inserted into potential mothers were used in research or discarded. But people who believe that life begins at conception consider this practice immoral and unacceptable.

Scientists have now found other sources for stem cells. Adult-derived stem cells from tissues that normally produce them (such as skin, blood, and umbilical cord blood) are effective for treating specific diseases.

Scientists are also now learning to make adult stem cells. They can reprogram cells by turning certain genes on or off. This can return the cells to their original undifferentiated state. Replacing embryonic stem cells with adult stem cells may remove the major moral objection to stem cell research.

been prokaryotic cells that entered eukaryotic cells and developed a symbiotic (mutually helpful) relationship with them. She presented evidence from many fields, including microscopy, genetics, molecular biology, and fossil and geological data, to support the theory. Her 1981 book *Symbiosis in Cell Evolution* described the impact of endosymbiosis on evolution.

More recent evidence from gene sequencing and analysis of evolutionary relationships of organisms further supports endosymbiosis. The DNA of chloroplasts and mitochondria is very similar to that of bacteria. Many genes from the original bacterial DNA now reside in the host cell's nucleus. Ribosomes (organelles that are sites of protein synthesis) in mitochondria and chloroplasts resemble bacterial rather than eukaryotic ribosomes. Both organelles undergo binary fission, just as bacteria do. Finally, scientists have recently found bacterial symbionts living inside modern eukaryotic cells—for example, in the guts of cockroaches and in protists.

The GENETIC CODE

The more scientists learned about genetic information in the cell, the more questions this new knowledge presented. After figuring out how genetic information was transferred to the next generation by meiosis, they wondered about the nature of the information contained in DNA. What did the information consist of, how was it transferred, and how did it lead to the development of a new organism, with all the characteristic structures and functions necessary to survive? In other words, they wanted to figure out the genetic code.

The DNA molecule was first isolated in 1869, but its structure was not described until 1953. The building block of DNA is the nucleotide, consisting of a phosphate group, a five-carbon sugar, and one of four nitrogen-containing bases. These bases include adenine (A), thymine (T), guanine (G), and cytosine (C). DNA consists of two twisted strands (a double helix) linked by paired bases, which bond together in a specific way (A to T and G to C). DNA size is described in terms of the number of base pairs in a chromosome or a genome. DNA in most species consists of many base pairs (billions in humans).

The nature of the genetic code remained a mystery until 1961. Scientists finally discovered that the genetic code consists of the arrangement of bases along the length of a single strand of DNA. Each set of three bases along the DNA strand acts as a code for one amino acid. Many amino acids, linked together in a specific order, form a protein. A gene is a length of DNA that contains sets of three bases in a specific order; this is the information needed to make one protein. The cell must be able to translate the sequence of the sets of bases into a unique arrangement of amino acids in a protein.

But DNA is located in the cell's nucleus, and proteins are made on organelles called ribosomes, found in the cytoplasm outside the nucleus. For protein synthesis to occur, the DNA information must be transcribed from DNA (which stays in the nucleus) to another nucleic acid, RNA, which can travel from the nucleus into the cytoplasm. Then, the genetic code in the RNA must be translated into a specific arrangement of amino acids. The processes of transcription and translation, working together, result in protein synthesis. Protein synthesis on ribosomes involves the interaction of several kinds of RNA. Every gene along an organism's DNA codes for a specific protein needed for that organism to function. Every cell in the organism's body is constantly making whatever proteins that cell needs.

Our understanding of the secrets of DNA and heredity are still being unraveled, but the twentieth century saw an explosion of work in this new field of molecular genetics. As the electron microscope made the details of DNA structure visible, it became possible to sequence DNA, or to determine the exact order of the bases along the DNA strands. This in turn made it possible to figure out the genome of any organism. DNA sequencing, or gene sequencing, has influenced areas in science and medicine. Many resources have been invested in the field of DNA sequencing. For example, the Human Genome Project, an international project to map all genes of the human species, was completed in 2003. In addition, the development of techniques for manipulating genes—moving genes from one species to another, for example—opened up the possibility (and the reality) of curing diseases, developing better agricultural products, and more.

GENETIC ENGINEERING

Genetic engineering, or genetic manipulation, is one of the key applied fields that has blossomed from the original cell theory and the developing field of molecular genetics. Understanding of gene structure and function quickly led to the question: How can we use this knowledge? The answer was, if we can sequence genes, we can also change them. If we can change genes, we can change how organisms look and function. We can even make new gene products. Genetic engineering manipulates a cell's DNA to change its genetic makeup or produce biological products.

Genetic engineering officially began in 1973, when Herbert Boyer and Stanley Cohen inserted a piece of DNA containing a gene that conferred tetracycline resistance into a bacterial cell that was not resistant. This technique is called gene splicing or recombinant DNA. In this technique, a desired gene from one organism is copied and reproduced so that many copies are available. These copies are spliced into another organism, often a bacterium. The bacteria can then be reproduced, and all new generations will have the new gene.

One of the first uses of genetic engineering was the development of a recombinant bacterium to produce insulin for diabetics. Scientists know the structure of the insulin gene because all human genes were sequenced during the Human Genome Project. They build the gene in the laboratory and insert it into the DNA of bacteria. Cultures of the recombinant insulin-producing bacteria are grown in huge vats. The insulin is harvested, purified, and sold as medicine. Before genetic engineering, insulin for diabetics was obtained from pigs and cattle in slaughterhouses.

Another important genetic engineering technique is the polymerase chain reaction (PCR), or DNA amplification. PCR reproduces a specific segment of DNA many times, making enough copies to be easily analyzed. These DNA fragments are used in DNA fingerprinting

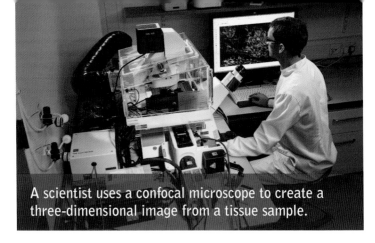

A scientist uses a confocal microscope to create a three-dimensional image from a tissue sample.

or profiling. This includes analyzing crime scene DNA, determining paternity of a child, or other situations requiring comparison of DNA samples.

Genetic engineering was a major breakthrough in cellular and molecular biology. It has resulted in the production of insulin and other drugs; bacteria that can break down oil spills and industrial waste; and new agricultural products such as disease- and insect-resistant crop plants, livestock that produce more milk or grow faster, and others. These genetically modified organisms (GMOs) have benefited people in many ways, but they have also raised many legal and ethical questions. Two of the many questions raised by this new technology are: When a new organism is created by genetic engineering, who owns it? What harm, if any, will GMO plants cause when released into natural ecosystems?

REPRODUCTIVE TECHNOLOGIES

Reproductive technologies use scientists' understanding of fertilization and early development to give hope to prospective parents who are having trouble conceiving. The simplest of these techniques is artificial insemination, in which a doctor simply inserts previously collected sperm into the woman's reproductive tract. This process may be used if the man's sperm count is low. Sperm from a number of ejaculations are collected and inserted at the same time, increasing the total sperm count.

Assisted reproductive technology (ART) involves the removal of unfertilized eggs from the woman's ovary, followed by replacement of either a fertilized egg or an egg and sperm mixture into the woman's reproductive tract. One type of ART is in vitro fertilization, in which unfertilized eggs are removed, fertilized outside the body, and grown into early embryos (blastocysts that have about 200 to 250 cells) before being inserted into the woman's uterus. The blastocyst implants in the uterus and grows to term. The first of these so-called test-tube babies was born in England in 1978. Since then, millions of babies around the world have been conceived by this technique.

In vitro fertilization is not always successful. Failure often occurs because the sperm is unable to penetrate the egg membrane. In newer methods, doctors use a needle to penetrate the egg cell and inject sperm directly into the egg cytoplasm. This procedure was first done in 1992. Thousands are now done in the United States every year, but only about 25 percent are successful. ART is still a work in progress and is one of the growing fields related to cellular and molecular biology.

The FUTURE of CELL BIOLOGY

What is the likely future of cell biology? Most likely the future of cell biology will be multifaceted, and the field will move forward in many directions. Today, cell biologists expect a laboratory equipped with genome-editing tools, a thermal cycler (also known as a PCR, or polymerase chain reaction, machine), and a fluorescent microscope. Rapid sequencing of an organism's genome is expected. Just a generation ago, in the 1990s, scientists had none of that. Cell biology was at a stage that now seems primitive. As a scientific field matures, it goes through predictable changes. One of these changes is improved technology.

In cell biology, the most obvious technology is microscopy. In the 1980s, modern electron microscopes and related technologies were added to typical light microscopy. These advances so far have included spinning disk, deconvolution, and two-photon microscopy, all new types of microscopy that improve precision, clarity, and accuracy when viewing tiny structures within cells. They allow scientists to observe cells and their contents at sizes as small as nanometers (billionths of a meter).

The viewing of these tiny structures, including molecules and macromolecular complexes (groups of very large molecules) within cells, is also being improved by the new technology of cryo-electron microscopy, or cryo-EM. In this method, specimens are viewed in a transmission electron microscope (TEM) under cryogenic (extremely cold) conditions. Cryo-EM itself involves many different techniques. Depending on the technique, scientists can view specimens, including intact tissue sections, individual bacteria and viruses, and protein molecules.

Various techniques can be combined to provide more comprehensive views.

Another predictable change has been the progression of cell biology from primarily descriptive (for example, describing the cell and its organelles) to studies that are more quantitative and more process-oriented. Some of these changes are the result of a new form of light microscopy—fluorescence imaging. In this technique, fluorescent dyes or proteins are used to label molecules or other structures in living cells. This allows viewers to follow processes such as gene or protein expression or other molecular actions in cells as they happen. This high-resolution imaging of living cells over a period of hours or days is a giant leap forward in understanding cell processes. Scientists are no longer limited to viewing dead cells, fixed at one point in time.

These changes in technology and techniques are being used to pursue many directions in both basic and applied cell biology. Other cell biologists are looking less toward advances in technologies and instead considering the broader implications of exactly what topics future cell biologists will be studying.

The editors of the *New Atlantis*, a technology and society journal, look forward to the increasing use of adult stem cells, those made by reprogramming adult cells to have the characteristics of embryonic cells. The technique is still being perfected, but their use would eliminate the ethical issues surrounding the use of embryonic cells. Moreover, scientists would have an unlimited source of cells for research. This is also a breakthrough for the applied field of regenerative medicine, which uses molecular or cellular engineering to replace or regenerate human cells, tissues, and organs. Stem cells could eventually be genetically reprogrammed, enabling people with serious conditions, such as diabetes, to regain their normal function.

Kai Simons, writing in the journal *Molecular Biology of the Cell*, points out that the machinery of the cell (its nanomachinery) has, on a miniature scale, all the components of an Earth factory. Cells can use energy, make products, and remove waste. They also transport and recycle materials efficiently. He sees a future in which understanding the tiny factories within the cell will help future humans overcome problems of dwindling natural resources and excess fossil fuel use.

Simons recommends a new direction in cell biology—the field of molecular bioengineering, in which cell biologists use engineering strategies to figure out how molecular processes within cells work. Engineering design principles would provide a framework for understanding how different classes of cellular molecules work—motor proteins or signaling proteins, for example. By understanding how life works on this cellular and molecular level, future engineers could apply this knowledge to the larger world, using it to find new ways to produce energy and materials on Earth.

The IMPACT of CELL THEORY

Scientists begin studies of natural systems such as cells because they are curious; they want to know how natural processes work. They do not know where their studies will lead because they are working in uncharted territory and uncovering new knowledge to the human species. Human understanding of living cells has progressed almost

Cryo-electron microscopy enables scientists to view the spherical Zika virus, shown here, in great detail.

unimaginably since the 1600s, when Hooke first observed dead cork cells. Because of the curiosity and inventiveness of previous generations, information about cells has continued to accumulate. Today's scientists can magnify cells hundreds of thousands of times and observe their processes at these great magnifications while they are still alive.

Cell theory has summarized the key evidence of our current knowledge. It is highly unlikely that any tenets of cell theory will be removed, but future breakthroughs may very well cause others to be added. Even if not, those future breakthroughs will improve human life, through treatments and cures for diseases and injuries, better agricultural products and methods, and perhaps even new products and energy sources. Whether or not humans are aware of cell scientists and what they do, cell theory has had, and continues to have, a profound effect on our everyday lives.

CELL THEORY
Chronology

1590 Zacharias Jansen invents the first compound microscope.

1665 Robert Hooke publishes his *Micrographia*, including views of sliced cork tissue under a microscope. Hooke also coins the word "cell" to describe the tiny chambers inside.

1668 Francesco Redi attempts to disprove spontaneous generation.

1674 Antonie van Leeuwenhoek is the first person to see living microscopic organisms, which he names "animalcules." Two years later he observes the first bacteria.

1802 Czech scientist Franz Bauer first describes the cell nucleus.

1825 François-Vincent Raspail first states the third tenet of cell theory: "Every cell is derived from a preexisting cell" (later credited to Rudolph Virchow).

1828 German chemist Friedrich Wöhler synthesizes urea, leading to the demise of vitalism.

1831 Robert Brown describes the nucleus of the orchid and coins the term "nucleus."

1832 Barthélemy Dumortier describes "binary fission" (later called cell division, or mitosis) in plants, leading him to reject spontaneous generation.

1837 Jan Purkinje states that animals are composed of cells and cell products (and possibly fibers) and that plant and animal tissues are similar.

1838 Matthias Schleiden concludes "all living plants are made of cells" and publishes a paper on this concept.

1839 Theodore Schwann concludes that all organisms are made of cells. He also publishes a cell theory with three tenets, the third of which is incorrect.

1842 Karl Wilhelm von Naegeli first discovers chromosomes, which he calls "transitory cytoblasts."

1852 Robert Remak publishes his observations of cell division and concludes that cells reproduce by binary fission (now mitosis).

1855   Rudolph Virchow concludes that all living cells come from preexisting cells. He publicizes (without credit) Remak's idea that cells reproduce by binary fission (now mitosis).

1856   Nathanael Pringsheim observes a sperm cell penetrate an egg cell.

1857   Rudolf Albert von Kölliker first sees mitochondria in cells, describing them as "granules."

1859   Louis Pasteur finally disproves spontaneous generation and gives credence to the third tenet of the cell theory that "cells come only from preexisting cells."

1869   DNA is isolated for the first time.

1879   Walther Flemming introduces the term "mitosis."

1883   Wilhelm Roux proposes that each chromosome carries a different set of heritable elements. Edouard van Beneden describes how chromosomes move during meiosis. Andreas Schimper distinguishes among types of plastids, including chloroplasts.

1890   August Weissman first explains the importance of meiosis in reproduction and heredity.

1898	Carl Benda names mitochondria.
1902	Walter Sutton and Theodor Boveri independently connect Mendel's "factors" in pea genetics with the behavior of chromosomes during meiosis, leading to the chromosome theory of heredity.
1931	The electron microscope is invented.
1953	James Watson and Francis Crick announce the structure of the genetic material, DNA.
1961	The genetic code is determined.
1970	Lynn Margulis publishes the theory of endosymbiosis.
1973	Herbert Boyer and Stanley Cohen begin the field of genetic engineering by splicing a new gene into a bacterium.
1978	The first test-tube baby (conceived by in vitro fertilization) is born in England.
2003	The Human Genome Project is completed.

CELL THEORY

Glossary

binary fission The process of asexual reproduction that occurs in bacteria, in which the DNA duplicates and the single cell then divides to form two new, identical daughter cells.

cellular respiration The chemical process by which cells break down chemical compounds (primarily glucose) to make adenosine triphosphate (ATP) and release energy to power life processes; occurs in all living cells.

chloroplast Plastid (pigment-containing organelle) found in plants and algae; its green pigment, chlorophyll, is necessary for photosynthesis.

chromosome A long segment of DNA, or genetic material, located in the nucleus of the cell; each species has a unique number of chromosomes, which contain the genes needed to define that species and its organisms.

DNA A type of organic compound (specifically, a nucleic acid) that contains the genetic information and is inherited, or passed from generation to generation, in all living organisms; the abbreviation for deoxyribonucleic acid.

electron microscope A microscope in which specimens are viewed by directing a beam of electrons through them.

embryology The study of embryos (formed when a sperm fertilizes an egg) and their development into living organisms; a person who studies embryos is an embryologist.

endosymbiotic theory The theory that chloroplasts and mitochondria, now organelles in eukaryotic cells, began as prokaryotic cells that were absorbed and became an interdependent part of eukaryotic cells.

eukaryotes Organisms having true cells, or large cells with membrane-bound nuclei and other complex organelles; includes all animals, plants, and fungi.

gamete Sex cell (egg or ovum in females, and sperm or spermatozoon in males); a male and a female gamete combine during sexual reproduction to form a new individual.

gene A piece of DNA, located on a chromosome; each gene provides instructions to the cell for how to make a specific protein, and all genes taken together provide a blueprint for all the structures and activities of the cell.

genetic code The set of rules defining how DNA carries and transfers genetic information, in which each set of three DNA nucleotides codes for a single amino acid, thereby determining the order of amino acids in a specific protein.

genome The complete set of genes present in every cell of every organism and species.

heredity The transfer of genetic information from cell to cell, organism to organism, and generation to generation.

histotechniques Techniques for preserving plant and animal tissues for viewing under a microscope; includes preserving the specimen to prevent decay, fixing it in paraffin, and slicing it with a microtome.

light microscope Also called compound microscope; a scientific instrument used to magnify an object held in place on a stage; light is shined through the object, and two or more lenses successively magnify it.

meiosis The type of cell division used to form gametes (eggs and sperm); the cell's chromosome number first decreases by half, and the new sex cells are then duplicated, resulting in four cells from the original single cell.

metabolic Having to do with the chemical reactions and changes occurring in living cells and tissues.

microtome An instrument used to cut materials (such as biological tissues) into extremely thin slices so they can be observed microscopically.

mitochondrion (plural **mitochondria**) The organelle in eukaryotic cells that is responsible for releasing energy for cell activities through cellular respiration; called the powerhouse of the cell.

mitosis The type of cell division that occurs in all cells except gametes and results in two identical daughter cells; responsible for growth, maintenance, and repair of body cells.

organelle A specialized structure within a living cell that carries out a specific function (such as reproduction or transport) for that cell; the nucleus and ribosome are examples.

pathology The science of the causes and effects of diseases, particularly the study of diseased tissues in the laboratory.

photosynthesis The process used by plants and algae to make organic food compounds, using energy from sunlight, carbon dioxide, and raw materials from air and water.

polymerase chain reaction (PCR) Also called DNA amplification; a genetic engineering technique in which many copies of a segment of DNA are reproduced, providing sufficient DNA for use in running comparison tests.

prokaryotes Unicellular organisms included in the kingdoms Bacteria and Archaea; their cells are extremely small and lack nuclei and other organelles.

spontaneous generation Also called abiogenesis; the now-disproven concept that living things can spontaneously occur from nonliving things, as maggots from dead meat or mice from wheat husks and sweaty clothes.

stem cell Rapidly growing, undifferentiated cell with the potential to differentiate into any kind of cell; the cells in very early embryos are stem cells.

vitalism A disproven scientific belief that assumed living material contained a "vital force" that was not present in nonliving material.

Further Information

BOOKS

Allen, John. *The Importance of Cell Theory*. San Diego, CA: Referencepoint Press, Inc., 2015.

Bond, Dave. *Genetic Engineering*. Broomall, PA: Mason Crest, 2017.

Cobb, Allan B. *Cell Theory.* Broomall, PA: Chelsea House Publishing, 2011.

Gibson, Karen Bush. *Cells: Experience Life at Its Tiniest*. White River Junction, VT: Nomad Press, 2017.

Schwebach, J. Reid. *Cell Structure and Function: Mastering the Big Ideas*. San Diego, CA: Cognella Academic Publishing, 2017.

WEBSITES

The Biology Project

http://www.biology.arizona.edu/DEFAULT.html

This interactive website from the University of Arizona offers resources and activities to help students learn about cells.

Khan Academy

https://www.khanacademy.org/science/biology

The biology page of the Khan Academy website leads to a wealth of basic information on cells and cell theory, much of it presented in video form, with quizzes.

National Institutes of Health (NIH)

https://www.nih.gov/research-training/science-education

This website includes various resources for students, including a microscope imaging station, fun tests, and interactive activities.

Texas Gateway: It's All About Cell Theory

https://www.texasgateway.org/resource/its-all-about-cell-theory

This website teaches about cells through interactive videos, handouts, questions and answers, and simple text.

CELL THEORY
Bibliography

Alberts, Bruce, et al. "The Chemical Components of a Cell." In *Molecular Biology of the Cell*. Bethesda, MD: National Center for Biotechnology Information, 2002. https://www.ncbi.nlm.nih.gov/books/NBK26883/.

Allen, Colin. "Teleological Notions in Biology." *Stanford Encyclopedia of Philosophy*, March 20, 1996. https://plato.stanford.edu/entries/teleology-biology/.

Andersen, Hanne, and Brian Hepburn. "Scientific Method." *Stanford Encyclopedia of Philosophy*, November 13, 2015. https://plato.stanford.edu/entries/scientific-method/#HisRevAriMil.

"Antonie van Leeuwenhoek." Famous Scientists. Retrieved February 1, 2018. https://www.famousscientists.org/antonie-van-leeuwenhoek/.

Arnold, Paul. "Famous Scientists: Walther Flemming and the Discovery of Mitosis." BrightHub, April 27, 2009. http://www.brighthub.com/science/genetics/articles/33094.aspx.

Blamire, John. "The Cell Theory." Brooklyn College, 2001. http://www.brooklyn.cuny.edu/bc/ahp/LAD/C5/C5_CellTheory.html.

Cook, Maria. "Modern Cell Theory." Sciencing, February 12, 2018. https://sciencing.com/modern-cell-theory-5492537.html.

Davidson, Michael W. "Marcello Malpighi (1628–1694)." Florida State University, November 13, 2015. https://micro.magnet.fsu.edu/optics/timeline/people/malpighi.html.

"Eukaryotic Cell vs. Prokaryotic Cell." Diffen. Retrieved March 5, 2018. https://www.diffen.com/difference/Eukaryotic_Cell_vs_Prokaryotic_Cell.

Eveleth, Rose. "There Are 37.2 Trillion Cells in Your Body." *Smithsonian*, October 24, 2013. https://www.smithsonianmag.com/smart-news/there-are-372-trillion-cells-in-your-body-4941473/.

"History of Biology: Cell Theory and Cell Structure." Biology Reference, 2018. http://www.biologyreference.com/Gr-Hi/History-of-Biology-Cell-Theory-and-Cell-Structure.html.

"History of Cell Biology." BioExplorer, August 17, 2016. https://www.bioexplorer.net/history_of_biology/cell-biology/.

Maaya, Inbar. "Theodor Heinrich Boveri (1862–1915)." *Embryo Project Encyclopedia*, March 3, 2011. https://embryo.asu.edu/pages/theodor-heinrich-boveri-1862-1915.

Mallery, Charles. "Cell Theory." University of Miami Department of Biology, February 11, 2008. http://fig.cox.miami.edu/~cmallery/150/unity/cell.text.htm.

Minai, Mandana. "Jan Evangelista Purkyne (1787–1869)." *Embryo Project Encyclopedia*, June 5, 2014. https://embryo.asu.edu/pages/jan-evangelista-purkyne-1787-1869.

"Modern Science: What's Changing?" University of California Museum of Paleontology, 2018. https://undsci.berkeley.edu/article/modern_science.

Phillips, Theresa. "Pros and Cons of Stem Cell Research." Balance, October 11, 2017. https://www.thebalance.com/pros-and-cons-of-stem-cell-research-375483.

Robinson, Richard. "History of Biology: Cell Theory and Cell Structure." Biology Reference, 2018. http://www.biologyreference.com/Gr-Hi/History-of-Biology-Cell-Theory-and-Cell-Structure.html.

Sadava, David, David M. Hillis, H. Craig Heller, and Sally D. Hacker. *Life: The Science of Biology*. Eleventh Edition. Sunderland, MA: Sinauer Associates, Inc., 2017.

Schultz, Myron. "Rudolph Virchow." *Emerging Infectious Diseases* 14 (2008): 1480–1481. https://www.ncbi.nlm.nih.gov/pmc/articles/PMC2603088/.

Schwann, Theodor. "Theory of the Cells." In *Microscopical Researches into the Accordance in the Structure and Growth of Animals and Plants*, translated by Henry Smith, 186–215. London: The Sydenham Society, 1847. http://mechanism.ucsd.edu/teaching/philbio/THEODOR%20SCHWANN.htm.

Shuttleworth, Martyn. "History of the Scientific Method." Explorable. Retrieved February 18, 2018. https://explorable.com/history-of-the-scientific-method.

Siegel, Andrew. "Ethics of Stem Cell Research." *Stanford Encyclopedia of Philosophy*, January 28, 2013. https://plato.stanford.edu/entries/stem-cells/.

Woodgett, Jim, and Danielle T. Loughlin. "Enabling the Next 25 Years of Cell Biology." *Trends in Cell Biology* 26 (2016): 789–791. http://www.cell.com/trends/cell-biology/fulltext/S0962-8924(16)30139-8.

Index

Page numbers in **boldface** are illustrations.

abiogenesis, 6
Altmann, Richard, 49, 71
Appert, Nicolas, 17–18
Aristotle, 5–6, 12

bacteria, 26, 41, 75–77, 80, 86, 103–104
Baer, Ernst Karl von, 38
Bauer, Franz, 40
Benda, Carl, 49, 69, 71
Berzelius, Jons, 70
Bichat, Xavier, 14
binary fission, 41, 43, 81
binomial nomenclature, 6
Boveri, Theodor, 48, 54, 63, 66–68, **68**
Brown, Robert, 40, 56

cell biology, future of, 106–108

cells
 animal, 34, 36–38, 40, 43, 46, 56, **58**, 58–59, 77
 blood cells, 26, **72**, **79**
 composition of, 88–90
 differentiation, 83, 98–99
 discovery and understanding of, 8, 25, 32–33, 51, 56, 71, 73, 75
 eukaryotic cells, **76**, 77, 81, 95–100
 formation of, 80–83
 number of in humans, 81
 plant cells, 34, 36–38, 40–41, **42**, 46, 50, 56, 58–59, 77
 prokaryotic cells, **76**, 77, 80, 95–100
 reproduction of, 40–48, 51, 54, 63, 79
 sex cells, 38, 44, 48, 66–67, 81, 84, 88, **97**

transformation of energy and, 40, 48–51, 63, 69, 79, 90–93
cell structure, 78–79
cell theory
 impact of, 108–109
 influence of, 95–109
 major players in, 53–71
 microscopes and, 29, 33, 53, 56
 modern, 73–75
 origins of, 9, 38, 41, 46–47, 53, 56–58, 60–61, 73
 tenets of, 9, 32–33, 36–37, 41, 46–47, 53–54, 59, 62–63, 65, 69, 71, 73–93, 109
cellular respiration, 36, 49, 69, 91–93, **92**, 95
chloroplasts, 50, 77, 90–91, 96, 100
chromosomes, 43–44, 48, 64–68, 81–82, 84–88, **85** 101
classification systems, 5–6, 75, 95
cork, 23, **24**, 25, 73, 109
cryo-electron microscopy, 106, **109**

Descartes, René, 7, 13

DNA, 44, 77, 80–83, 89, 96, 100–104
Dumortier, Barthélemy, **30**, 41
Dutrochet, Henri, 36–37

egg cells, 38, 44, 48, 66, 81, 84, 88, **97**, 105
endosymbiotic theory, 96–100
eukaryotic cells, **76**, 77, 81, 95–100

Flemming, Walther, 43, 54, 63–67, **64**, 81

Galen, **10**, 13, 62
genetic code, 96, 100–102
genetic engineering, 96, 102–104
genome, 83, 101–103, 106
Greeks, ancient, 5, 7, 9, 11–14
Grew, Nehemiah, 33–34

Haeckel, Ernst, 50, 69
heredity, 44, 48, 65, 67–68, 71, 74, 84, 102
Hippocrates, 11–13
Hooke, Robert, 22–26, 29, 33–34, 40, 73, 86, 109
 image of microscope of, **23**

Index 125

humors, 7, 12–13, 51, 62, 75

Jansen, Zacharias and Hans, 21, 24

kingdoms, 75–78, 95
Kölliker, Rudolf Albert von, 48–49, 69, 71

Leeuwenhoek, Antonie van, 21–25, 29, 33, 40, 86
image of microscope of, **27**
Liebig, Justus von, 54
life, early ideas about, 5–9, 11–14
Linnaeus, Carolus, 6, 56

Malpighi, Marcello, 33–34
Margulis, Lynn, 96–100
meiosis, 44–48, 51, 66–68, 74, 84, **85**
Mendel, Gregor, 48, 67
Micrographia, 25–26
microscopes, 21–29, 37, 41, 54, 96, 106
cell theory and, 29, 33, 53, 56
compound, 21–25, **23**, 86
confocal, **104**
cryo-electron microscopy, 106, **109**
electron, **72**, 84, 86–87, 90, 102, 106
evolution of, 86–87
Hooke's, 22–25, **23**, 86
invention of, 8, 17, 21
inverted, **28**
Leeuwenhoek's, 22–26, **27**, 86
light, 86, **94**, 106–107
preparing specimens for, 34–35
scientific method and, 20
Microscopical Researches into the Accordance in the Structure and Growth of Animals and Plants, 58, 60
microtome, 8, 34–35, **35**
mitochondria, 48–50, 69, 71, 77, 79, 86, 91, 96–100
mitosis, 41–43, 47, 51, 64–65, 74, 80–84, **82**
Müller, Johannes, 57, 61

Naegeli, Karl Wilhelm von, 43
Needham, John, 15–18, 20
nucleus, 40, 44, 56, 58, 64–65, 79

organelles of, 38–40, 48, 69, 71, 74, 77–78, 90, 95, 100

Pasteur, Louis, 18, 20, 54, 57
photosynthesis, 36, 50, 69, 71, 90–93, **92**, 95–96
plagiarism, 46–47
plastids, 48, 50, 69, 71
Plato, 9, 12
prokaryotic cells, **76**, 77, 80, 95–100
Purkinje, Jan, 37–38, **39**, 46

Raspail, François-Vincent, 41, 43, 47, 62
red blood cells, 26, **79**
Redi, Francesco, 15, **16**, 17–18, 20
Remak, Robert, 43
reproductive technology, 96, 104–105
Roux, Wilhelm, 43, 54, 63, 65–67

Sachs, Julius, 50, 69, 71
S-bend flask (Pasteur), 18, **19**, 20
Schimper, Andreas, 50, 69, 71
Schleiden, Matthias, 41–43, 46, 54–59, **55**
Schwann, Theodor, 38, 41, **45**, 46, 54, 57–59, 60–61
  drawings by, **58**

scientific experimentation, 9, 15–18, 20, 31, 51
scientific theory, 31–33
Spallanzani, Lazzaro, 17–18, 20
spermatozoa/sperm, 26, 44, 48, 66, 81, 84, 88, 104–105
spontaneous generation, 6–8, 14–18, 20, 29, 33, 43, 51, 57, 59, 80
stem cell research, 96, 98–99, 107–108
Sutton, Walter, 48, 54, 63, 67–69

Trembley, Abraham, 41
Tyndall, John, 18

van Beneden, Edouard, 67
Virchow, Rudolf, 41, 43, 46–47, 54, 59–63, **63**
vitalism/vital force, 7, 13–15, 17, 36–37, 51, 57, 61, 70, 75

white blood cell, **72**, **79**
Wöhler, Friedrich, 14, **52**, 54, 69–70

Zika virus, **109**

About the Author

Carol Hand has a PhD in zoology from the University of Georgia. Hand has taught college biology, written biology assessments for national assessment companies, and written middle and high school science curricula for a national company. For the past nine years, she has been a freelance science writer. She has authored more than fifty science books, including biology titles such as *Introduction to Genetics*, *Epidemiology*, and *Vaccines* as well as environmental titles, including *Bringing Back Our Oceans* and *Climate Change: Our Warming Earth*. She lives in Kansas.